Challenges of Human Space

Springer
London
Berlin
Heidelberg
New York
Barcelona
Hong Kong
Milan
Paris
Santa Clara
Singapore
Tokyo

Marsha Freeman

Challenges of Human Space Exploration

Published in association with
Praxis Publishing
Chichester, UK

Marsha Freeman
Associate Editor
21st Century Science & Technology
P O Box 16285
Washington DC
USA

SPRINGER–PRAXIS BOOKS IN ASTRONOMY AND SPACE SCIENCES
SUBJECT *ADVISORY EDITOR*: John Mason B.Sc., Ph.D.

ISBN 1-85233-201-8 Springer-Verlag Berlin Heidelberg New York

British Library Cataloguing-in-Publication Data
Freeman, Marsha
Challenges of human space exploration. – (Springer–Praxis books in astronomy and space sciences)
1. Outer space – Exploration 2. Outer space – Exploration – History 3. Space biology
I. Title
629.4

Library of Congress Cataloging-in-Publication Data
Freeman, Marsha
Challenges of human space exploration / Marsha Freeman.
p. cm. – (Springer–Praxis books in astronomy and space sciences)
Includes bibliographical references.
ISBN 1-85233-201-8 (alk. paper)
1. Manned space flight. 2. Outer space–Exploration. I. Title. II. Books In.

TL873.F74 2000
629.45–dc21 00-020843

Apart from any fair dealing for the purposes of research or private study, or criticism or review, as permitted under the Copyright, Designs and Patents Act 1988, this publication may only be reproduced, stored or transmitted, in any form or by any means, with the prior permission in writing of the publishers, or in the case of reprographic reproduction in accordance with the terms of licences issued by the Copyright Licensing Agency. Enquiries concerning reproduction outside those terms should be sent to the publishers.

© Praxis Publishing Ltd, Chichester, UK, 2000
Printed by MPG Books Ltd, Bodmin, Cornwall, UK

The use of general descriptive names, registered names, trademarks, etc. in this publication does not imply, even in the absence of a specific statement, that such names are exempt from the relevant protective laws and regulations and therefore free for general use.

Copy editing and graphics processing: R. A. Marriott
Cover design: Jim Wilkie
Typesetting: BookEns Ltd, Royston, Herts., UK

Printed on acid-free paper supplied by Precision Publishing Papers Ltd, UK

This book is dedicated to the astronauts and cosmonauts who lived on board Mir; to the Russian and American space agency managers who demonstrated the courage necessary to complete the Shuttle–Mir programme; and to the late Congressman George E. Brown Jr, whose reasoned and unwaivering support for the US space programme for two decades provided leadership for past and future accomplishments in space exploration

Table of contents

Foreword

We know from earliest recorded history, some 5,000 years ago, that human beings have always sought to learn more about their world. The century just ended has witnessed stunning advancements in science and medicine, including the launching of space exploration. There is a danger, however, that the new century may usher in an age of timidity, in which fear of risks and the obsession with cost-benefit analysis will dull the spirit of creativity and the sense of adventure from which new knowledge springs.

As a physician and surgeon, I am a researcher, an explorer, a participant in today's quest for knowledge. Medicine, after all, is about exploration of the human body. Medical research in the space exploration programme and in the International Space Station may provide keys to some of the most vexing problems that affect human health today.

The unique microgravity environment of space can be used to investigate ways to treat or prevent deteriorating physical conditions that affect the elderly and disabled, which are mimicked in healthy astronauts during space missions. A space station provides a facility, unavailable on Earth, to observe these processes and to develop countermeasures that may apply not only to astronauts but also to the aged and feeble. The space station is a necessity, not a luxury, any more than a medical research center at Baylor College of Medicine is a luxury.

Present technology on the Space Shuttle allows for periods of only about two weeks in space, but we do not limit medical researchers to a few hours in the laboratory and expect cures for cancer. The joint missions between NASA and the Russian Space Agency that permitted American astronauts to live and conduct experiments onboard the Mir space station for months at a time gave up a glimpse of what can be accomplished during much longer missions in orbit. The laboratories of the next international station will provide researchers with the most sophisticated experimental medical equipment, which may lead to new knowledge and even breakthroughs.

Concrete examples of technologies awaiting long-term research in space demonstrate the benefits a space station holds for medicine. Two that are discussed in depth in this book are tissue modelling and protein crystal growth.

Producing exact replicas of human tissues in the NASA-developed rotating wall vessel, or bioreactor, promises important insights for cancer research, organ transplant research, and human virus culturing. Space-grown protein crystals which are usually larger, more developed and more uniform than those grown on Earth – allow more exact analysis of their molecular and cellular structures. Such analysis can then be used to design and test specific treatments for diseases.

In 1984 I performed a heart transplant on a NASA engineer, after which, fortunately, he was able to resume his activities at the Johnson Space Center. He became interested in our experimental laboratory research on the artificial heart, and this led to my collaboration with NASA engineers and a formal joint NASA/Baylor project to develop a miniature ventricular assist pump that would fulfill all of the physiologic and biocompatability requirements for supplementary cardiac output. The result was the development of a miniaturised axial flow pump, about the size of an AA battery, that meets the stated objectives of our design strategy. It has now been used as a bridge to transplant in more than twenty patients in Europe, with gratifying results.

In general, the space programme's requirements for miniaturised and highly reliable instrumentation and sensors were precursors of cardiac pacemakers. The need to monitor astronauts' vital signs while hundreds of miles away from Earth has led to medical telemetry for monitoring hospital patients' vital signs. The same telemetry has permitted paramedics to save the lives of countless patients while en route to hospitals. And space telemetry has spurred the development of telemedicine, which allows clinical consultations and support in underserved and disaster-stricken areas worldwide.

All these advances are the serendipitous result of scientific and technological research. We often investigate to uncover questions we do not yet know how to ask, and to discover answers we never expected.

Space exploration is human exploration. The knowledge we gain in space comes not only from sending people beyond Earth, but also from marshalling human resources on Earth that make spaceflight possible.

Our space programme is a symbol to the rest of the world that the United States looks to the future and plans to maintain its leadership role in science, technology and research. It demonstrates that our leaders have the vision to look beyond today's challenges and are committed to continue improving the human condition.

We conduct research not only to find the right answers, but also to ask the right questions. In the history of science and technology development, the greatest advances have been made by those who wondered why and sought to discover how. This is why we go into space. That is why we explore. And that is the genius of humanity.

Michael E. DeBakey, MD

Chancellor Emeritus; Olga Keith Wiess and Distinguished Service Professor, Michael E. DeBakey Department of Surgery; Director, DeBakey Heart Center, Baylor College of Medicine, Houston, Texas

20 January 2000

Author's preface

During the past decade, literature about the space programme has become increasingly dominated by a pessimism that is entirely antithetical to the content and spirit of the space programme itself. Rather than chronicle the accomplishments and overcoming of adversity by explorers in space, authors of articles in newspapers and magazines, and writers of books, focus inordinate amounts of attention upon failures, crises, delays and cost overruns.

An example is Bryan Burrough's book *Dragonfly: NASA and the Crisis Aboard Mir*, published in 1998. Mr Burrough documents the events in excruciating detail, and focuses almost exclusively upon the lack of trust between American and Russian participants at the start of the Shuttle–Mir programme, the personality flaws of some of the astronauts and cosmonauts who lived rather unhappily on Mir, the lack of attention by NASA mission managers to the needs of the first astronauts on Mir, and the seriousness of the fire on the Mir and collision with the unmanned Progress cargo ship in 1997. It was the publication of that book that prompted the writing of this current work.

Problems have certainly occurred in space, as will many more in the future, but the disturbing nature of many current accounts of space exploration is that they report on the problems and failures, with no apparent appreciation of the difficulty of the task being attempted, and little appreciation for the skill and courage of the space explorers or the commitment and dedication of the support and management personnel.

To grasp the enormity of the task of opening new frontiers, and the courage and sacrifice made by the explorers who conquered them, it is instructive to reflect on the expeditions that were launched to the last frontiers on Earth.

In 1937, Charles R. Keg published his book *The Story of Twentieth Century Exploration*. In his preface, Key observes: 'The lure of the unexplored has a universal magnetic attraction... Some, with vaunting ambition, delve down to the fathomless depths of the ocean, or soar into the stratosphere and dream of the day when excursions to the Moon or into stellar space will be practicable. The lure of the unexplored will be as potent a thousand years from now as it was when old Pytheas set off from Massilia to discover Britain for the Greeks.'

Jack Stuster's *Bold Endeavors: Lessons from Polar and Space Exploration*, published in 1996, presents extensive documentation of the expeditions to the South Pole. The following account is taken primarily from these two sources.

In 1907, Ernest Shackleton announced his intention to reach the South Pole and find the exact position of the magnetic pole. Charles Key describes this adventure as comparable to 'setting off from the east coast of England across a frozen North Sea, and they had not the faintest idea of what the southern shores were like, or even if there was a southern shore'. Early the following year, Shackleton and his small team set off on their adventure. On 9 January 1909 they reached their farthest point south, at a latitude of more than 88° S. But their food supplies had run out, and they had no choice but to turn back.

After the South Pole had been reached by Roald Amundsen in 1911, Shackleton decided to organise the Imperial Trans–Antarctic Expedition of 1914–1915, to further explore the Antarctic. It was reportedly a particularly cold winter that year, Jack Stuster writes, and their ship *Endurance* became trapped in pack ice before reaching the continent. For five months the 28 members of the expedition lived on the ship, until it was finally crushed by the pressure of the ice, and sank. For another five months the crew camped on a floating piece of ice. 'As the men listened in the darkness' of the polar winter, Stuster recounts, 'killer whales cruised the edges of the floes and bumped them from beneath in attempts to knock their prey into the frigid sea'.

With each storm that pounded the sea and ice, the floe on which Shackleton's crew was encamped became smaller and smaller, and they eventually had to abandon the ice when it began to break up. Finally, after 497 days, they reached Elephant Island, with the three small boats they had saved from *Endurance*, but they could not possibly use them to make the return trip home.

Shackleton set off with five members of the crew to reach South Georgia, and get help. After 497 days in a trapped ship and on ice floes, Shackleton embarked upon what Jack Stuster describes as 'one of the most remarkable passages in the history of seafaring'. The small boat, of about 24 feet, was batted around for 16 days of hurricane–type conditions, while the crew suffered from frostbite and intermittent seasickness. When they finally reached South Georgia they were on the opposite side of the island from the whaling station, which was their destination. Rather than risk another bout with the sea, they chose to cross the island on foot. This had never been done before, and was not accomplished again for another 40 years. The explorers knew that it was 'only' 29 miles to the other side of the island, but they were faced with crossing glaciers and a snow–covered mountain range. Two men were too weak to traverse the island, so a party of three set off on what Stuster calls a 'fantastic adventure'.

The three men climbed the cliffs, crossed the glaciers, nearly froze and, after 37 hours, reached the whaling station. The whalers had assumed that all of the crew on *Endurance* had perished months before. More than four months after he had left Elephant Island and the remainder of the crew, Shackleton returned to take them all home. The entire crew (not one man was lost) returned to England, after having been 'castaways' for nearly two years.

While the adventures of seven American astronauts on Mir pale in comparison to the voyages of Ernest Shackleton, the Shuttle–Mir programme had its exciting moments. When the programme was proposed, Mir was already substantially past its five–year rated lifetime. The programme itself was a political initiative to cast aside the 40 years of the Cold War and begin the process of returning to a normal relationship with the former Soviet Union. NASA decided it was worth the risk to send American astronauts to the ageing Mir station, because the experience that would be gained from 20 years of long–duration spaceflight that the Russians had accumulated would be invaluable to the International Space Station.

Virtually anything that could (and did) happen on Mir could also happen on the new station. NASA has learned valuable lessons from the crises on Mir, and these are being used to increase the safety and productivity of the astronauts and cosmonauts who will work together throughout the next decade on the International Space Station.

No one would suggest that people should take on bold challenges and incur great risk for no reason, but there are no better reasons than the furthering of human knowledge, the uplifting of the human spirit, and the increase in mankind's ability to shape nature, on Earth and throughout the Solar System.

Acknowledgements

I am extremely grateful to the astronauts, scientists and space programme managers who were very generous with their time for interviews and in answering questions. I would especially like to thank Dr Michael E. DeBakey, the reknowned heart surgeon, scientist, inventor, medical statesman and humanitarian, for contributing his inspiring Foreword. This book has also greatly benefited from the encouragement of my colleague Dr Philip Harris, and the enthusiasm of Clive Horwood, Chairman of Praxis. Photographs and diagrams are reproduced courtesy of NASA, except where otherwise indicated.

List of illustrations

Chapter 1

Chapter 2

Chapter 3

Chapter 4

Chapter 5

Chapter 7

1

Skylab sets the stage

From the earliest time that man believed it would be possible to explore space, plans were being laid to not only go there, but to work in this new environment above the Earth. In 1903 – the year that the Wright brothers proved that man could don wings and soar above the surface of this planet – Russian space pioneer Konstantin Tsiolkovsky completed his work on the basic scientific principles required for space flight. In his book *Beyond the Planet Earth* he included a concept of establishing a space station as an 'outpost in the wilderness', leading to orbiting settlements that would pave the way for expeditions to the Moon, Mars and the asteroids.

In 1923, German scientist Hermann Oberth's small work *The Rocket into Planetary Space* provided the physical concepts and mathematical detail to begin the design, testing, and eventual manufacture of vehicles to take man to Earth orbit. In the final section of his book, Oberth proposes the construction of 'observation stations' in orbit. While he observes that there may seem to be no reason to take men into space, 'often enough, things which seemed to have no relation to daily life, later achieved the highest practical value (I am reminded here of electricity).' The observation station, he proposes, could become a permanent 'miniature moon.' It could allow for communications between remote regions of the globe, allow the recognition of details on Earth for military and geographic purposes, observe icebergs and warn ships at sea, and provide light, through the use of orbital mirrors, for nighttime illumination, and to influence weather and climate. Finally, Oberth states that 'the observation station could also serve as a fuelling station' which would be in place around the Earth, and later other planets, to enable more efficient forays to planetary surfaces.

Hermann Oberth's student, Wernher von Braun, led the German effort that proved man could build a machine that would take him outside the Earth's atmosphere. In 1942 this feat was accomplished with the first successful launch of the V2 rocket. Based on that accomplishment, von Braun and his colleagues knew that after the war, space would be opened for mankind. To garner popular support for such an effort in the US, in 1952 von Braun authored a series of articles for the magazine *Colliers* – later complied into books – which included the concept of a huge space station in the shape of a wheel, which would rotate to produce artificial gravity, 1,075 miles above the Earth.

The early space pioneers, who had laid out a reasoned and incremental approach to explore the planets, assumed that before man ventured far from home he would build Oberth's 'observation stations' to learn how to live and work in space. However, when President John F. Kennedy announced on 25 May 1961 that the United States, within the decade of the 1960s, would 'land a man on the Moon, and return him safely to Earth', plans for the space station were necessarily put on the back burner. To meet the President's deadline there would be no time to proceed with the intermediate step of a space station.

But while engineers and mission planners were still trying to determine how to even put a man on the Moon, the post-Apollo Moon programme was already taking shape on the drawing boards. It was assumed that projects such as space stations would follow the Apollo programme, which would be just the first step in the exploration of space. In 1952 von Braun had written: 'Development of the space station is as inevitable as the rising of the Sun; man has already poked his nose into space and he is not likely to pull it back.'

In 1963 NASA's Office of Manned Space Flight began to develop options for programmes to follow Apollo. The Mercury, Gemini, and Apollo programmes would demonstrate that man could safely reach Earth orbit, travel to the Moon, and return. For man to continue to explore more distant objects, however, he would have to be able to *live* in the microgravity environment of space for months, if not years. The space station was the next logical step.

By 1965 the massive Saturn V rocket that would take astronauts to the Moon had accomplished its first successful (unmanned) test, increasing NASA's confidence that President Kennedy's goal would be reached. That year, the Apollo Applications Office was opened to oversee programmes that would make use of the capabilities developed for the lunar landing. In 1969, when Neil Armstrong, Buzz Aldrin, and Mike Collins returned from the first landing on the Moon, an initially very ambitious post-Apollo programme had been considerably pared down, to became Skylab.

1.1 A HOUSE IN SPACE

The final design for the first space project that would be explicitly designed to conduct scientific experiments relied heavily on the heritage of the Apollo programme. The workhorse of the complex would be the Orbital Workshop, to be outfitted with the equipment and storage of supplies required for months of work in space, as well as the accommodations for each three-man crew. The basic structure of the workshop was a cylindrical container, 14.7 metres long and 6.5 metres wide, which was the S-IVB third stage of the Saturn V rocket that took astronauts to the Moon.

The Orbital Workshop was divided into two floors, to allow a separation between various activities. On the bottom floor were the equivalent of the bedrooms, kitchen and dining rooms, bathroom, and exercise facility, as well as the experiments that the astronauts would perform on themselves to chart their adaptation to weightlessness.

Figure 1.1. America's first space station, Skylab, at the end of its mission in 1974. Notice the missing solar array, and a heat shield patch on the front of the spacecraft.

The upper floor contained lockers for the storage of consumables, experiments, a supply of water, vaults to hold film, and the astronaut manoeuvring unit that would be the precursor to the Manned Manoeuvring Unit that NASA would test on the Space Shuttle many years later.

Attached to the Orbital Workshop was the Airlock Module, allowing crew-members to exit the space station to perform routine as well as emergency extravehicular activities (EVAs). The Airlock Module could be evacuated for egress or ingress by a space-suited crew-member, who would travel to open space through an EVA hatch on the side. Mounted on the truss frame of the Airlock Module were the oxygen and nitrogen storage tanks required for the operation of Skylab's life support system. Inside were the electric power control and distribution station, environmental control system, communications, and data handling and recording systems – a kind of basement 'nerve centre' for the station.

The Multiple Docking Adapter (MDA) attached to the Airlock Module provided the portals to allow the Command–Service Module bringing the crew to Skylab to dock with and enter the station. Inside the MDA were a film vault, spare parts and a window. A number of experiments were mounted in the MDA, including the Earth Resources Experiment Package, a materials processing facility, and the control and

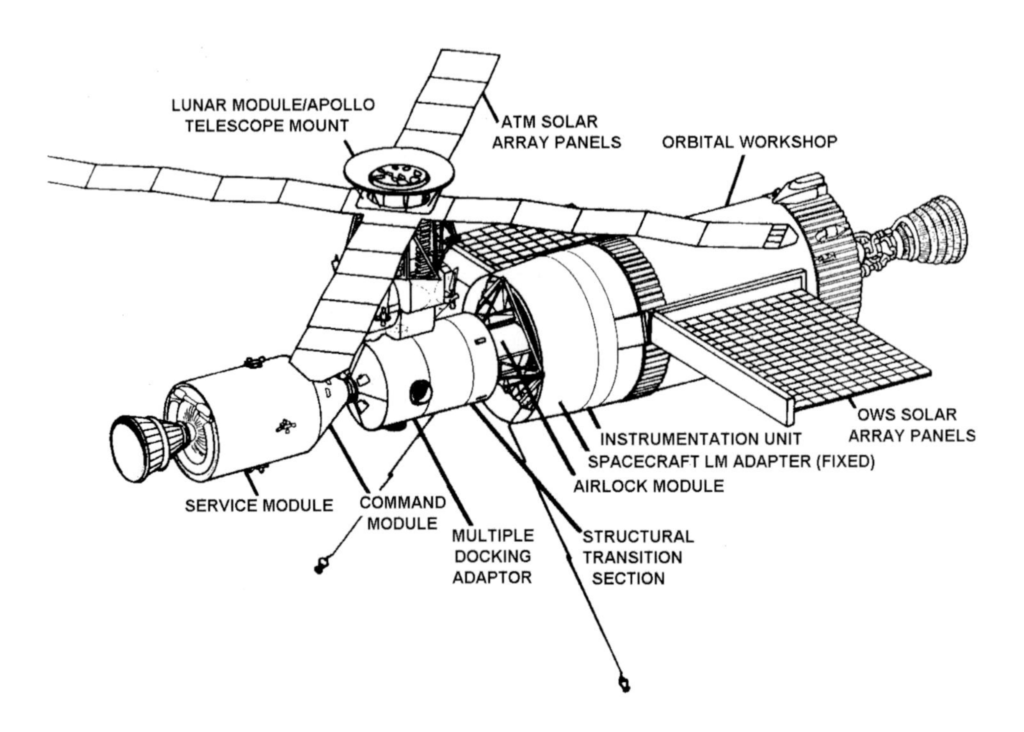

Figure 1.2. By 1968 the basic design of Skylab had been committed to paper. It was launched unmanned into Earth orbit on 14 May 1973.

display console for the magnificent Apollo Telescope Mount. The normal operation of the docking system was to have the Command–Service Module, piloted by the crew, dock at the forward end of the MDA. In case of emergency, however, a side docking port could have been used by a second Command–Service Module, had it been needed for a space rescue.

The Docking Adapter served another important function as the mechanical support for the cluster of eight telescopes designed to observe the Sun – the Apollo Telescope Mount (ATM). The project originated from the Apollo programme, when scientists had hoped to place the complex telescope package on the Lunar Module, where it could observe the Sun over a wide spectrum from visible light to X-rays. The ATM would provide physicists and astronomers with a new and breathtaking view of the Sun.

When the Command–Service module was docked with Skylab and became part of the cluster, it had a volume equivalent to a three-bedroom house, weighed almost 100 tons, and was as tall as a 21-storey building. For the three crews of astronauts who would live there for 34, 59 and 84 days, Skylab was, indeed, their 'house in space'.

1.2 MAN AS SUBJECT

While Skylab would provide scientists with unprecedented views of the Earth and the heavens, the prime subject of the experiments aboard the laboratory was man himself. If he were to be able to travel to more distant ports than the Moon, it would have to be proven to the satisfaction of the doctors responsible for his health and well-being that exposure to microgravity would not prevent his return to Earth without lasting deleterious effects.

At the same time, it was necessary to evaluate the performance of men while they were in space. There would be no point in sending them there, were they only to be too sick or disabled to accomplish productive work. In a 1969 article describing Skylab, Werner von Braun stated: 'The heavy work schedule prepared for this observatory will also furnish valuable lessons about human proficiency for difficult scientific work performed under zero-gravity conditions over an extended period of time.'

The interest of the medical professional in this new venue for man's exploration was summed up by Richard Johnston, Director of Life Sciences at the NASA Johnson Space Center in Houston, at the Skylab Life Sciences Symposium held in August 1974, after completion of the third and final Skylab mission. 'History is filled with examples of man's desire to explore new frontiers,' he stated. 'Having sensed the thrill of discovery, man has pressed on to scale new heights, not weighing the cost or personal risk, but mindful only of his destiny to conquer the unknown. He crossed the seas in open boats and the wastes of the arctic on dog sleds. He perished but rose again until there were no longer any new seas to cross or mountains to climb, or arctic poles to visit. He had explored his Earth.'

Dr Johnston noted that while exploration has always been a 'risky undertaking', it is the case that 'the preservation of life and health is essential to the successful conquest of the unknown... Man's opportunity to explore is largely dependent upon the advancement of technologies in transportation and life support.' For the first time, he said, explorers were sent abroad with the express purpose of conducting studies on themselves, to not only see if they could endure, but to document their responses to their new environment.

Project Mercury, the first American manned space programme, which orbited single astronauts for as long as 34 hours, dispelled many basic concerns about the frailty of the human being upon being exposed to space flight. Predictions that men would become disoriented, be unable to perform even simple mechanical functions, or even fall unconscious, proved to be unfounded. The two-man Gemini missions, which extended man's stay in space to 14 days, further demonstrated that man could perform useful tasks, adapt to the weightless environment, and re-enter the Earth's atmosphere and readapt to Earth's gravity. But even over a two-week period, significant changes were observed in human physiology, which Skylab would investigate for longer periods of up to nearly three months.

During the 171 days that Skylab contained human occupants, hundreds of medical measurements were made. Skylab crewmen were monitored closely for signs of the first affliction that could be produced by microgravity – motion sickness, or

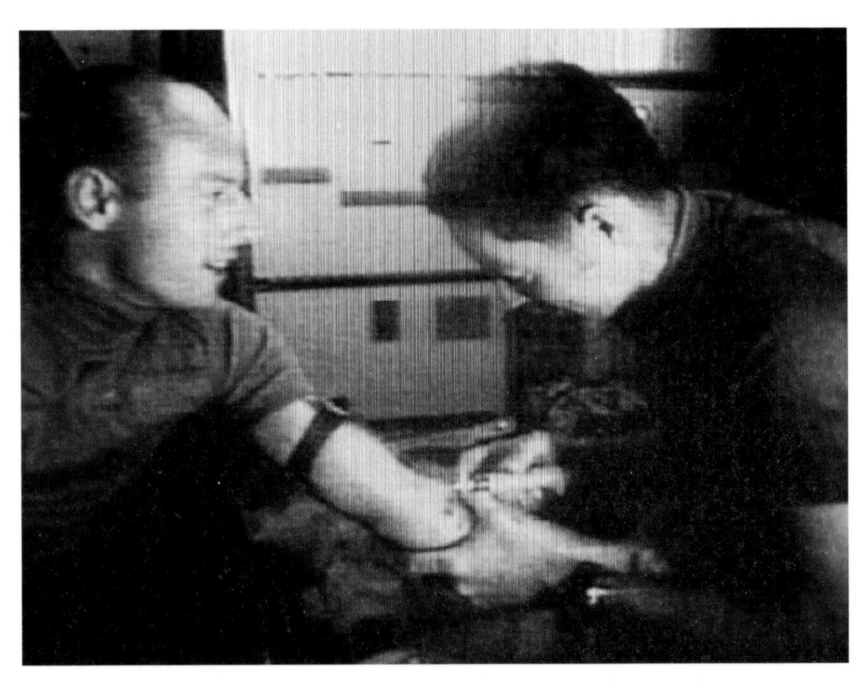

Figure 1.3. Medical data collected during the three Skylab missions rewrote the textbooks on human physiology. Here, Joseph Kerwin (right), a medical doctor, and the first Skylab mission science pilot, draws blood from Charles Conrad Jr, the mission commander, on 6 June 1973. The blood sample was part of the Haematology and Immunology Experiment.

space adaptation syndrome. While this malady had not been prevalent during previous American manned space missions, it had been noted in more than half the Soviet crew-members. During the three Skylab increments, five of the nine astronauts *did* experience motion sickness, with one suffering symptoms even into his fourth day of the mission.

Medical specialists have been not only concerned about the discomfort that this syndrome causes the crew-member, but the effect it could have on his ability to carry out tasks, which, in the first few hours of a mission, could be critical for the safety of the crew and the entire success of the flight. Of special interest was that during the first Skylab mission, none of the astronauts suffered from motion sickness. This has been attributed to the fact that their first hours in space were dedicated to an extraordinary effort to repair the damage which the laboratory had sustained on launch, and a resulting delayed encounter with the cavernous space of the workshop.

According to a report in 1983 contained in the groundbreaking NASA publication *Space Physiology and Medicine*, 'Based on the subjective reports of the three Skylab crews, and vestibular experiments conducted during the flight, it was concluded that space motion sickness remains a problem, is not predictable by the usual Earth-bound tests, and can be alleviated somewhat by the prophylactic administration of medications. The search for the optimum medication and schedule of administration continues.' Indeed, it continues to the present day.

Earlier short-duration space missions had indicated consistent changes in the cardiovascular functioning of crew-members. While this adaptation to reduced stress on the heart was certainly lawful under conditions where the heart had considerably less work to do than under the influence of the one-gravity environment of Earth, medical concerns were whether or not such changes continued unabated throughout a long-duration mission, or reached a self-limiting point, as the answer to that question could have serious consequences for astronauts returning to Earth. On Skylab, doctors chose to observe the condition of orthostatic intolerance caused by the redistribution of fluid away from the head upon return to Earth, and the electrical activity and changes in the size of the heart. They found that the changes appeared to be self-limiting, and would stabilise over a period of four to six weeks, with no apparent impairment of crew health or performance.

It was already known from short spaceflights that there is bone mineral loss and mineral imbalance as a consequence of the microgravity environment. Skylab metabolic studies of the crew showed a significant increase in the excretion of urinary calcium, which continued throughout the period of the flight, and was more noticeable in the lower, rather than the upper, extremities. There was no evidence that this bone thinning – similar in many ways to the osteoporosis suffered by patients on Earth – was self-limiting, as it continued steadily throughout the entire mission. Similarly, significant loss of nitrogen and phosphorus also occurred, presumably caused by the loss of muscle mass and of fluid.

Overall, long-term space missions received a fairly clean bill of health from the medical professionals attending to the Skylab astronauts. While dramatic changes are endured by the human physiology in its adjustment to space, no critical effect was found to be irreversible over time, following return to Earth.

1.3 ON STRIKE!

Perhaps the most unexpected and dramatic medical problems encountered during the Skylab programme were the emotional and psychological effects of the isolation of spaceflight, observed, in particular, during the third, 84-day Skylab mission. All three crew-members were first-time rookies in space, experiencing their first encounter with microgravity and what they sometimes perceived as inconveniences, as compared to Earth. The mission began badly, when all three men endured a bout of space motion sickness; but worse than the stomach queasiness and discomfort was the decision of the crew to disregard NASA regulations.

When pilot William Pogue vomited, the astronauts should have informed Mission Control, and they should have freeze-dried the product and brought it back for post-flight analysis. But they did neither. Unbeknown to the crew, their conversation deciding to disobey orders was being recorded, and would later be heard on the ground. As Henry Cooper reports in his book on Skylab, it was the first time astronauts were ever reprimanded in public during a flight. This early incident created a tension between the crew and Mission Control which worsened as time went on, and took the ground nearly half the mission to try to set right.

From the illness they suffered, their unfamiliarity with microgravity, and their irritation with Mission Control, the crew started to fall further and further behind in their work schedule. Mission control made the critical mistake of expecting that the third Skylab crew would be a repeat of the energetic and enthusiastic second crew, and while the third crew fell further behind in the timeline, flight directors and mission managers were trying to find yet more experiments for the crew to carry out, lest they complete everything ahead of schedule and become idle. The minute-by-minute micro-managing of the crew's schedule from the ground kept up a constant pressure, increasingly being seen by the astronauts as unreasonable.

Cooper reproduces in his book one commentary by crewman Pogue, which located the problem, but in a tone never before heard by Mission Control: 'You have to put away equipment, you have to debrief, and then you have to move from one position to another, and you have to look and see what's coming up, and we're just being driven to the wall!' In another instance during the third Skylab mission, he remarked, 'There's not enough consideration given for moving from one point in the spacecraft to another and allowing for transition from one experiment to another... When we're pressed bodily from one point in the spacecraft to another with no time for mental preparation, let along getting the experiment ready, there's no way we can do a professional job! Now, I don't like being put in an incredible position where I'm taking somebody's expensive equipment and threshing about wildly with it and trying to act like a one-armed paperhanger trying to get it started in insufficient time!'

The pace of experimental work being demanded by Mission Control reached the point that the crew had virtually no time off, and hardly a meal that was not interrupted with more instructions from the ground. Commander Gerald Carr tried to remind Mission Control that 'on the ground, I don't think we would be expected to work a sixteen-hour day for 85 days, and so I really don't see why we should even try to do it up here.' As flight surgeon Dr Jerry Hordinsky said after the mission, the medical specialists believed Mission Control was pushing the crew too hard, but had little say regarding the flight schedule, which was considered a 'non-medical duty.'

Finally, two things became clear to Mission Control. First, they had been pushing the crew too hard to try to accomplish too many tasks in too little time. This was proven to be the case during the post-flight analysis of video and audio recordings made during the three Skylab missions. The study revealed that in space it took considerably longer to complete tasks than it did during pre-flight training on Earth. For the third Skylab crew, that figure was 58% longer – comparable with the second flight's in-flight time of 54% longer than on Earth. The pressure from Mission Control had only exacerbated the situation, leading to mistakes, and the crew falling further and further behind. Second, and most disturbing, the crew's interaction with the ground was characterised more and more by complaining and griping, and less and less by problem solving or trying to find constructive ways to correct the situation.

At the end of their sixth week aboard Skylab, the third crew went on strike. Commander Carr, science pilot Edward Gibson, and Pogue stopped working, and spent the day doing what they wanted to do. As have almost all astronauts before and after them, they took the most pleasure and relaxation from looking out the window at the Earth, taking a lot of photographs. Gibson monitored the changing activity of the Sun, which had also been a favourite pastime of the crew.

Commander Carr made the crew's concerns known to the ground: 'We'd all kind of hoped before the mission, and everybody had the message, that we did not plan to operate at the second crew's pace. And I think really up here, my biggest concern is keeping the three of us alert and healthy. I think an illness is probably something that we can really do without.'

Carr and the crew seemed aware that the people on the ground with whom they spoke were under their own pressure from NASA management, and the scientists with experiments on Skylab. 'I'm also getting the feeling from some of the questions that have been asked of us the last few days, that people there are beginning to hassle over who gets our time and how much of it... We're beginning to get just little hints and indications that we're getting into a time bind – that it's got people really worried down there... And I'd like to know just exactly what everybody's motives are when they're asking these questions...'

With the first indication of conciliation on the part of the crew, Carr continued, 'I imagine you guys are probably caught right in the middle of it, and the questions that arise in my mind is, are we behind, and if so, how far? What can we do it we're running behind and we need to get caught up? What can we do that's reasonable? ... and we'd like to be in on the conversation, and we'd like to have some straight words on just what the situation is right now. Commander out.'

With the air cleared, and the concerns of the crew stated straightforwardly for the first time during the mission, the situation aboard Skylab improved markedly. Later, Cooper reports, Commander Carr blamed himself for not having earlier voiced the concerns of the crew. 'We had told the people on the ground before we left that we were not going to allow ourselves to be rushed; yet we got up here and we let ourselves just get driven right into the ground.'

But why did the crew take to bitching and complaining to Mission Control for six weeks, before finally having a constructive 'heart-to-heart' talk with the people they knew to be, in the end, most concerned about both their success and their welfare? Was this simply an American phenomenon; a strike for rugged individualism?

1.4 THE STRESS OF ISOLATED ENVIRONMENTS

While there appears to have been inadequate preparation for either the Skylab astronauts or the ground crews for possible psychological problems, especially between the crew and Mission Control, during extended space flights, Soviet medical personnel were quite attuned to the likelihood of stress and strain during long-duration flights. By 1970 the Soviet Union was planning not for just one temporary

Figure 1.4. The strain of living in the isolated environment of space was most noticeable in the crew's strained communications with the ground. This is the Mission Operations Control Room in the Mission Control Center at the Johnson Space Center in Houston. Skylab is visible on the screen at the top left.

space station, but a string of such increasingly complex facilities, making this not a theoretical but an operational question.

At the Third Orbital International Laboratory Symposium, held in Germany in October 1970, Professor O.G. Gazenko, from the USSR, reported that although the topic of psychological selection of compatible crews for space stations was still a little 'obscure', experiments were being conducted in the Soviet Union to study people under isolated conditions, 'such as those of Antarctic expeditions, and other such similar groups.' Observations were also being conducted on international undersea expeditions, as well as in specially outfitted laboratories on *terra firma*, to gain an insight into psychological changes that may be expected from the experience of long-term isolation in space.

The preliminary conclusions that Gazenko presented included the observation that groups become more effective if each member's ability to function in a group is taken into account, and that the previous experience of the group's members 'played a considerable role in this.' However, he stated, the characteristics of the individual's ability to function well in a group 'were not stable, and during the course of the experiment they tended to change. The more complex the conditions of the experiment, and the longer the experiment goes on, the greater are the dynamics that take place within the course of the experiment.' Although this insight did not make it possible for cosmonauts to completely avoid psychological stresses in orbit, it does appear to have prepared Soviet medical personnel to anticipate such problems, and to develop sets of support activities to try to ameliorate the effects.

Over the course of the 1970s the Soviet Union orbited progressively more complex and capable space stations. The launch of Salyut 6 in September 1977 began the era of long-duration orbital space missions that have lasted through the activities on the Mir. As cosmonauts were beginning to spend several months in space, Soviet medical specialists would be able to observe, and respond to, changes in the behaviour of the men they sent into space.

By the early 1980s the Soviets had made detailed studies of psychological stresses in space, with the longest visit to Salyut 6 having been six months. As recounted in *Space Physiology and Medicine* (second edition), they noted five general phases of task performance during extended flights. For the first few days, they observed, the crew-members adapt to microgravity and become familiarised with the station, with occasional intervention needed by ground control. For the first 10–15 days of the flight, they achieve stable and efficient performance, as major physiological functions have adequately adapted to weightlessness. After this optimal period there is a period of fatigue, but it is compensated for by the crew's high motivation to perform well. High tensions are seen from high workloads. The medical personnel note a following 'unstable compensatory period', with increasing periods of fatigue and decreased work capacity. There is evidence of emotional instability, with periodic sleep disturbance. In this period, changes are highly individual, and task performance is affected. Then, two or three days before return to Earth, the crew is generally in a high emotional state, and able to work at high efficiency levels.

The privations suffered by long-duration cosmonauts have been personally told by Valentin Lebedev, who was on board Salyut 7 from May to December 1982. In his book *Diary of a Cosmonaut*, Lebedev explains that on the day of his launch, 13 May, he took an oath. He promised himself that he would 'always remember' in any difficult situation, to 'follow my head, not my heart. I won't speak or act hastily.' He promised himself he would be sensitive to his crewmate, and keep his 'self control'. He recognised that 'the success of the mission depends on us, and only by the work we do will they judge me as a cosmonaut and as a man.'

About ten days after launch, Lebedev noted that he was not sleeping well, tossing and turning and thinking about his family and friends At the same time, he notes, they had been falling behind the timeline, and were staying up, sometimes until 2.00 am, to try to catch up. On 30 May, the cosmonaut notes in his diary that talking with his family via television made him sad. 'It's hard to believe that somewhere people are walking, smelling the flowers, swimming, and sunbathing... Today for the first time, I felt quite depressed.' By 3 June, Lebedev's growing irritation with the ground is noticeable, and he complains about the fact that the doctors worry about their health, but as soon as they have a day off, they are pestering them about medical experiments. And on 14 June, Lebedev self-consciously reports: 'On board, our relations are good, but our relations with ground control are becoming more complicated. It is probably not easy for them to feel our problems from down there.

Tomorrow we have to repair the sextant. I have to control myself more; I was bitching too much today during our communication period.'

By the end of June, Lebedev begins to tell ground control of the problems they are having in keeping up such a fantastic pace, and an agreement is made that the crew's schedule and the priority of experiments will be discussed by all concerned parties. But that does not solve every problem, as the ground still has to see that the science program is carried out, even as the cosmonauts suffer from fatigue and boredom. On 12 September, Lebedev advises that 'future programmes for space settlement and long-term flights should pay attention to... social–psychological problems, such as communications between people on Earth and those working in space.'

One lesson learned from the long-duration flights on the Salyut stations was that during the mission there should be a programme to boost morale and reduce the workload as the mission becomes more difficult psychologically. During Lebedev's 211-day Salyut 7 flight, the cosmonauts' work day was reduced from 16 hours to 13 hours during the last stages. During a 326-day flight on Mir, the work day was reduced from 8.5 hours to 6.5 hours, and then further reduced to 4.5 hours near the end of the mission.

Early in the Soviet long-duration programme, provision was made for cosmonauts to receive letters from home and favourite foods among the supplies delivered by unmanned Progress spacecraft. Crew-members had frequent two-way voice and video communication with family and friends, famous celebrities, and research colleagues, to try to ease the tension, depression, monotony and fatigue. As more experience was gained, more adjustments were made throughout the Mir programme. While no one would claim that there were not psycho-social problems in the future, current Russian space programme managers have many years of experience and varied tools at their disposal to try to make long-duration spaceflight happier and more productive for the crew.

A year after Valentin Lebedev returned to Earth, the crew of Space Shuttle flight 9 blasted off for Earth orbit, carrying more than 70 experiments in the first flight of the European-built Spacelab laboratory. Scientists and managers, eager to make the best use of the 10-day flight, pressed the crew to carry out many things simultaneously. As payload specialist Robert Parker was working on one investigation, Mission Control called to ask him to do something else. 'You guys need to recognise there are two people up here trying to get all your stuff done,' Parker told the ground. 'I think you might be quiet until we get one or the other done.' After a reply from the capsule communicator (another astronaut – the only person who talks to the astronauts from Mission Control), Parker continued: 'Would you guys please tell us exactly what you want done when and we'll forget about what we were doing at the present time...?' Why would seasoned and professional space travellers react with such a lack of patience and civility to their colleagues on the ground?

1.5 FRUSTRATED? KICK THE CAT!

Dr Nick Kanas, with the Department of Psychiatry at the University of California, San Francisco, who has carried out extensive studies on the psychological effects of isolation in space and during simulations on the ground, describes the behaviour of 'taking it out' on people on the ground as a displacement of emotions. At the same time that there may have been justifiable reasons for crew-members to be irritated with Mission Control in many different circumstances, it is also often the case that crew-members were frustrated at their own inability to carry out their tasks as they had planned, whether due to illness, the breakdown of equipment, or just the extra time required to find and set up an experiment.

As stated by Connors *et al.* in 1985: 'Unfulfilled expectations might have contributed to the Skylab 4/Mission Control conflict... Frustrations and disappointments which are relatively minor under normal conditions tend to be exaggerated under conditions of isolation and confinement. The mission planners, as the agents perceived responsible for the unpleasant state of affairs, were the likely targets for subsequent aggression.' Researchers have drawn such conclusions not only from studying long-duration spaceflight, but the behaviour of groups of people in many situations of isolation and confinement in a variety of hostile environments.

Each crew-member, knowing full well that he was going to spend months in close quarters, knew that for the mission's success he could not easily express any frustration or irritation with his crewmates, but all had to 'get along.' On Earth, if unpleasant interactions take place at work, one can always go home and 'kick the cat' to vent one's anger or frustrations. In space, where this was not possible, the next best thing was to blame Mission Control, the astronauts' only link to the outside world.

In a 1998 paper for the publication *Aviation, Space, and Environmental Medicine*, Dr Kanas surveyed the psychiatric problems that have afflicted participants of long-duration space missions, from both the United States and the Soviet Union. His concern is that both the support personnel on the Earth and the crew-members on the mission be prepared to deal with such difficulties, since an affected individual can have a 'devastating effect on the welfare of the crew and the ultimate success of the mission.'

Dr Kanas reports on the syndrome which the Russians have identified as effecting a large number of cosmonauts during long-duration missions: a deprivation caused by hypostimulation, which they call asthenia. The observable features are 'nervous or mental weakness manifesting itself in tiredness... and quick loss of strength, low sensation threshold, extremely unstable moods, and sleep disturbance.' Dr Kanas agrees that this syndrome is indeed prevalent, and should lead to the development of measures to prevent and counter the effects of asthenia, beginning with the support programme that is used in the Russian space programme.

In an earlier paper, published in 1993, Dr Kanas and Alan Kelly specifically studied the communication between space crews and ground personnel, on the basis of questionnaires that were distributed to 112 active and retired American astronauts, eight international astronauts from six Western countries, and eight

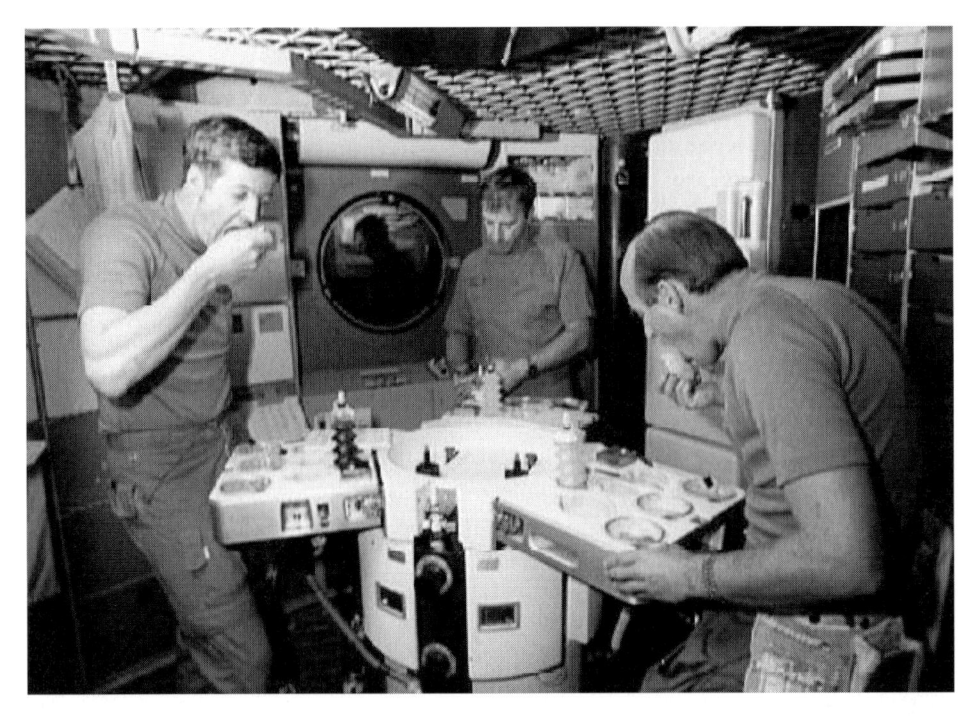

Figure 1.5. One of the most important activities for maintaining the cohesion of a space-crew is the opportunity to eat meals together, but when the workload became too hectic during long-duration missions, this critical social activity was often sacrificed in order to complete assigned tasks. Here, in the Skylab Orbital Workshop trainer in March 1973, the first Skylab crew dines on food prepared for their mission.

Soviet cosmonauts. 54 (42%) of them completed and returned the questionnaire. There was agreement that if there were shared experiences and a shared excitement of spaceflight between the crew and the ground, this improved communications between the two groups. More among the long-duration travellers (the Russians) than the short-duration Space Shuttle crews, the ability to be in contact with loved ones on the ground was judged to be important for the success of the mission.

What Skylab and the Soviet long-duration missions on their Salyut stations demonstrated is that for the relationship between Mission Control and a space crew to be productive and maintain a healthy psychological equilibrium, there has to be a recognition by the crew and the training personnel that demonstrable patterns of counter-productive behaviour in spaceflight, and similar isolated hostile environments, should be anticipated, and countermeasures put into place. As we shall see later, it was a lesson that NASA had not fully learned when the next opportunity for long-duration missions arose, 20 years after Skylab, in the Shuttle–Mir programme of the 1990s.

1.6 A SCIENTIFIC CORNUCOPIA

By any measure, regardless of the problems, Skylab's 171 days of crewed operations were a stunning success for the scientific community. Even those who had been skeptical that having men in space would produce results otherwise unobtainable, became converts.

The results from the Apollo Telescope Mount – including eight solar flares photographed by the astronauts during the Sun's quiescent period – had been completely unanticipated. The Principal Investigator for the ATM, Dr Richard Tousey, of the Naval Research Laboratory, remarked after the Skylab missions: 'The solar observations made by the ATM were extraordinarily valuable, perfect, and complete. In spite of innumerable problems, far more than ever imagined was accomplished. The solar observations retrieved are staggering in quality and quantity. Best estimates made by each [Principal Investigator] are that no less than five years of work by competent and sizeable teams are required to reduce and interpret the data, and ten years may well be needed.' Skylab rewrote the book on solar physics.

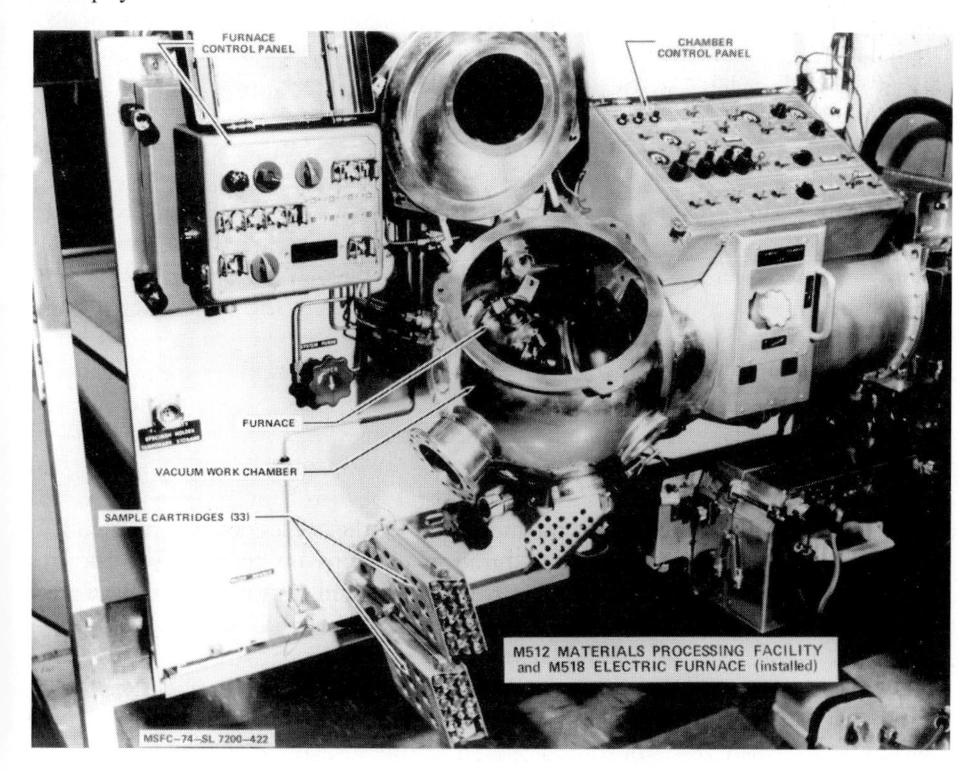

Figure 1.6. The Materials Processing Facility was mounted in the docking adapter of the Skylab space station. The astronauts conducted metallurgical experiments in the vacuum chamber. The Multipurpose Electric Furnace was utilised to observe the phase changes in materials processed at temperatures up to 1000° C.

Through the Earth observation programme, new sensors and instruments were tested, which would later be incorporated into unmanned remote sensing satellites. Geological and agricultural features were mapped, and astronauts were deployed to photograph momentary phenomena such as volcanoes. More than 46,000 photographs and 40 miles of electronically recorded data for the contiguous 48 states of the United States and other nations were returned from Skylab.

One series of experiments, to take advantage of the microgravity environment of space in the materials sciences, would have long-lasting results, and in the 1980s would spawn a new field of space applications that have the potential to revolutionise the care of human health and the battle against disease.

1.7 A PERFECT CRYSTAL

Recently, *Microgravity News*, published by NASA's Marshall Space Flight Center in Huntsville, Alabama, retold the story of how the space agency and the scientific community became interested in growing crystals in space. In the late 1960s, MSFC Director, Wernher von Braun, was concerned about producing high-quality spherical ball-bearings for precision guidance systems for satellites and spacecraft, and approached materials scientists at the Massachusetts Institute of Technology (MIT), to determine whether they thought that producing them in microgravity would allow perfectly spherical products.

While the MIT scientists, Gus Witt and Harry Gatos, did not believe von Braun's idea would bear fruit, they began to consider what types of materials research might benefit from the microgravity of space. 'We had just concluded that one of the primary complications in optimising properties of semiconductors is the obstacle of gravitational interference,' Gatos recalls. 'In a gravitational field, the melt, which we want to transform into a perfect solid, is subject to turbulance convection, which drastically interferes with the orderly incorporation of atoms into a solid. Some of these atoms are responsible for the electrical properties of the materials,' he explained, 'which effect the capabilities of semiconductors.'

Witt and Gatos designed semiconductor crystal growth experiments that were conducted on Skylab and also on the 1975 Apollo–Soyuz Test Project. 'We found that the moment you remove the primary perturbation, the convection caused by gravity, you saw other deficiencies that nobody was paying much attention to.' The MIT team was most interested in crystals that had 'defects' built into them, in the form of impurities which, when introduced, produced properties in the material that are critical for electrical conductivity and other characteristics. In space, the scientists found that these impurities are distributed more evenly than on the ground, leading to a higher performance of the material. Being able to see the difference between Earth- and space-produced materials allow scientists to understand the basic underlying physical mechanisms involved in crystal growth.

In the proceedings of the Third Space Processing Symposium to discuss Skylab results in May 1974, the paper presented by Dr Witt and the MIT team states: 'The results obtained prove the advantageous conditions provided by outer space. Thus,

Figure 1.7. Professor Harry Gatos, of the Massachusetts Institute of Technology, designed a Skylab experiment to observe the formation of crystals that were partially melted and then resolidified in microgravity. This crystal of indium antimonide was smoother and more perfect than those grown on Earth.

fundamental data on solidification thought to be unattainable because of gravity-induced interference on Earth are now within reach.' Stressing that the fundamental knowledge obtained in space would cause scientists to rethink not only theory but also their experiments on Earth, the MIT team stated: 'On the basis of the present results, it is no longer a matter of speculation that fundamental data necessary for bridging the gap between theory and experiment can be reliably obtained in the absence of gravity and that outer space presents one of the greatest opportunities ever afforded science and technology.'

Scientists had long conjectured that in crystal formation, the forces of convection, caused by temperature differentials, and of sedimentation, where a heavier material precipitates out of solution and falls to the bottom of the container, had placed limits on the size and perfection of crystals grown on Earth.

In one experiment, designed by H.U. Walter of the University of Alabama, in Huntsville, and carried out on Skylab, indium antimonide crystals brought from Earth were melted and then solidified while freely suspended from a seed crystal inside the electric furnace. The results from the experiment were stunning. The crystals grown in microgravity were single crystals, with a surface that was exceedingly smooth. When the crystals were examined on Earth, they indicated a higher degree of purity, homogeneity, and freedom from lattice defects than any Earth-grown crystals of similar composition. One crystal of germanium selenide, grown for Dr H. Wiedemeir of Rensselaer Polytechnic Institute, was 20 mm in length – larger than any previously grown on Earth. Overall, Dr Ernst Stuhlinger, from MSFC, described the Skylab results as 'extremely encouraging' in his summary presentation at the 1974 Space Processing Symposium.

When the materials processing experiments for Skylab were designed, it was expected that more perfect crystals would eventually be grown in factories in space, to eliminate the yearly loss of millions of dollars, due to defects from semiconductor manufacturing processes in Earth's gravity. Instead, it has been found that the improved understanding of the fundamental physics involved in crystal growth has enabled significant improvements in ground-based processing. In addition, it is still today too expensive to gain access to space for commercial production plants to be economical.

Nonetheless, the returns in basic science of crystal growth that resulted from Skylab encouraged the next generation of crystal experiments – this time with organic proteins – which has become one of the most promising fields for microgravity research, with enormous potential benefit for the fight against disease on Earth.

1.8 SPINNING A WEB IN SPACE

As scientists from dozens of disciplines were preparing their experiments for Skylab, to observe how everything from man to plants, insects, and materials would respond to microgravity, Martin Marietta official Ken Timmons proposed that tomorrow's scientists – that is, high school students – should also design experiments for Skylab. Such 'hands-on' experience, and close work with scientists and space engineers, he believed, would increase students' interest in science and technology.

In October 1971, in cooperation with NASA, the National Science Teachers Association mailed out 100,000 announcements soliciting experiment proposals to science teachers. By early the following year, more than 55,000 teachers had requested materials for entry, and 3,409 proposals were finally submitted, involving 4,000 students from all 50 US states. In March 1972, 25 national winners were selected, and 19 experiments later flew on the space station.

The student experiments ranged from studying the X-ray emission from Jupiter to studying the behaviour of bacteria in space. One of the most popular experiments was proposed by Judith S. Miles from Lexington High School, Massachusetts. Titled 'Web Formation in Zero Gravity', the experiment, which flew on the second Skylab mission, was to study the adaptation to microgravity by the common cross spider (*araneus diadematus*), by observing its attempt to spin a web. The Research Division of the North Carolina Department of Mental Health carried out the same experiment on the Earth, as a control.

This remarkably simple experiment produced some of the most stunning results from Skylab. The spiders, Anita and Arabella, which on Earth depended upon gravitational cues to spin their symmetrical webs, initially produced a jumbled mass of filaments. But over the course of their stay in space these tiny creatures were able

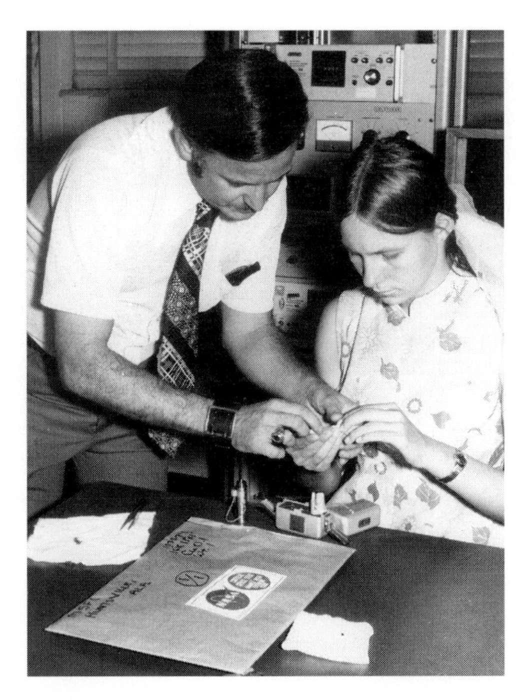

Figure 1.8. A critical aspect of carrying out scientific experiments in space is to involve young scientists on the ground to increase overall interest in science and technology among students. Through the Skylab programme to fly student experiments, Judith Miles, seen here with Dr Raymond Gause, her NASA Science Adviser, designed the Web Formation in Zero Gravity Experiment. In this picture the experimenters are opening the package that contains web material spun by spiders in space.

Figure 1.9. During her first attempts to spin a web without the benefit of gravity, Arabella – one of Judith Miles' spiders – did not succeed in producing any with the symmetry characteristic of spider webs on Earth. The web pictured here is clearly irregular.

Figure 1.10. By the third week, the spider Arabella had somehow determined a method by which the symmetry of the spacing between the radials, the angles between them, and the number of spirals from the centre could be created without gravity. Scientists inferred that there were sensory, motor and neurosensory changes that occurred in the spider during her adjustment to life without gravity.

to learn how to adjust to their peculiar circumstance and resume their normal activity, as seen in the photographs of Arabella's webs (Figures 1.9 and 1.10), taken three weeks into the mission.

1.9 THE LEGACY OF SKYLAB

The magnificent laboratory that had hosted nine men, and dozens of science experiments that would rewrite textbooks, and provided a window to what human explorers could do in space, was to suffer a humiliating end. When the laboratory was first designed it was hoped that the reusable manned Space Shuttle transportation system would be up and running by the late 1970s, when Skylab's orbit had decayed to the point that it needed a reboost to stay aloft. But continuous cutbacks and redesigns of the Shuttle system delayed its first flight until April 1981 – too late to save Skylab, which descended to Earth two years earlier.

Original plans for the post-Apollo programme envisioned two Skylab stations, and a second, which was perfectly able to fly, now sits among the artefacts of the space programme in the National Air and Space Museum in Washington DC.

While Skylab met an untimely end (and there would be a hiatus of six years in American manned spaceflights until the first Space Shuttle flight), its legacy stands untarnished in the history of space science and engineering. In addition to the operational lessons learned, and the confidence that man could live and work productively in space, Skylab created the teams of scientists who would continue this

Figure 1.11. Skylab succeeded in its mission, proving that man could live and work in space for months at a time; and Commander Gerald Carr demonstrated in this photograph taken on 1 February 1974, during the last Skylab mission, that it could also be fun.

new field of space science in the Space Shuttle era, on the Russian Mir station, and, in the future, on the International Space Station.

2

Americans and Russians on Mir

Throughout the late 1970s and into the early 1980s, while the United States had no manned flights into space, and then only short forays with the new Space Shuttle, the Soviet Union continued to improve its ability to house, employ, and service cosmonauts in orbit. By the time the Salyut 7 station was reaching the end of its useful life in 1986, the Soviets had demonstrated that man could productively live and work in space for more than half a year, that a system of unmanned cargo ships could be periodically docked with a station to replenish the supplies that are required for long-duration missions, that orbital stations could be repaired and maintained by a well-trained crew, and that with exercise and other prophylactic measures, man's health would not be permanently impaired after long encounters with weightlessness.

When the first module, or base block, of the Mir station was launched into orbit on 20 February 1986, the event was hardly noticed in the US. Only one month earlier, the Space Shuttle *Challenger* had exploded 73 seconds after launch, killing all seven crew-members. Much of the nation was still in a state of shock and mourning. But in the Soviet Union, a new phase of manned space activity was beginning.

Experts expected that the new station would see improvements over the later Salyut series, but few could imagine the complexity of what was under construction. Mir was the first modular space station, which, when completed, would be the largest complex of spacecraft ever to exist in space. It was meant to be permanently occupied.

The core module weighed 20 tons, and was designed to function as a habitation module and control centre. A pair of solar arrays provided 9 kW of electric power, mainly for life support and station-keeping. Seven computers in an integrated complex in the module, known as Strela, provided attitude control, relieving the cosmonauts from having to attend to that important function. Indicating the intended expansion of the new station, the Mir base block had six docking ports, for the simultaneous accommodation of two Salyut manned vehicles and the addition of more modules (see Figure 2.2).

2.1 FIRST LABORATORIES IN ORBIT

The second module, Kvant 1, was launched to Mir a little more than a year later, on 31 March 1987. It was an astrophysics laboratory, which also carried equipment for attitude control and life support. After it was attached to the core module, Kvant took over as the station's aft port, which received the unmanned Progress cargo ships.

Half the scientific payload of the Kvant laboratory, in terms of weight, was an electrophoresis system for processing biological materials. The other 800 kilograms consisted of a variety of astronomical instruments to measure high-energy emissions from space – Röntgen. The Max Planck Institute and the University of Tübingen in Germany built a high-energy X-ray experiment, as did the Netherlands Space Research Organisation, with Birmingham University. The Soviets supplied an X-ray specrometer, and an ultraviolet telescope was built by the Byurakan Astrophysical Observatory in Armenia. The European Space Agency also contributed scientific equipment for Roentgen.

As David Harland describes in his comprehensive book *The Mir Space Station*, Kvant 1 also supplied new technology to supplement the life support systems in the core module, including the Elektron apparatus to process reclaimed water to resupply oxygen to the atmosphere, and the Vozdukh hardware to scrub the cabin's carbon dioxide. As cosmonauts and astronauts would later discover, this redundancy in life support systems would provide a critical margin of safety under crisis conditions onboard Mir.

During Kvant's docking with the Mir, when it was at a distance of 200 metres from the complex, its thrusters failed to slow it down, and it flew past the station. Flight controllers on the ground considered aborting the mission, but decided to make a second docking attempt rather than abandon the laboratory. On the second try, a soft docking was effected, in which the docking unit penetrated the Mir docking unit but then became stuck.

An unscheduled spacewalk was therefore carried out from Mir, after cosmonauts on the ground rehearsed the planned extra-vehicular activity (EVA) in the neutral buoyancy tank at the cosmonaut training centre, which simulates microgravity. It was found that there was debris attached to the docking port. The cosmonauts freed the obstruction, and ground controllers completed the hard docking. In April, when a Progress supply ship docked with Mir, the Soviets were operating the first station composed of four separate spacecraft, including the Soyuz, docked at the station to provide the crew's transportation back to Earth.

The next module that was added to Mir was Kvant 2, launched on 26 November 1989. It carried an EVA airlock to allow cosmonauts to work on the outside of the Mir complex, two solar arrays, and science and life support equipment, intended to enhance the capabilities for the future station configuration. Additional oxygen generation systems, as well as hygienic facilities for the crew, were onboard Kvant 2. The module also included a broad range of scientific equipment, including an apparatus to study fluid flow, as Harland reports.

Also on board were Inkubator-2 which would be used to hatch eggs in order to

Figure 2.1. The Shuttle–Mir programme marked the beginning of a new era in manned space exploration. This photograph of the two spacecraft docked together was taken by the Mir 19 crew on 4 July 1995, as cosmonauts Anatoly Solovyov and Nikolai Budarin circled the spacecraft in their undocked Soyuz capsule.

study the growth of embryos, a multispectral and a mapping camera for Earth science studies, and astronomical instruments, which were mounted on a new platform that could be put into a scan mode using remote control.

The Kristall module was launched to the Mir on 31 May 1990. As its name implies, the research performed in Kristall was to focus on materials science experiments, utilising five furnaces. The laboratory also had an electrophoresis unit to complement facilities onboard Kvant 1, and the Svet greenhouse, which was to become a major delight for the cosmonauts, and later astronauts, onboard Mir. Kristall also had two stowable solar arrays for additional power, needed especially for the energy-intensive furnaces, and a docking port equipped with a special androgynous docking mechanism designed to receive spacecraft weighing up to about 100 tons. This docking unit was originally designed for the Soviet Buran shuttle, and was used by the Space Shuttle *Atlantis* on the first docking mission.

Following the launch of Kristall there was a five-year hiatus in completing the Mir complex. Dramatic political changes had taken place. In the economic chaos that was to result from the dissolution of the Soviet Union, and the uncertainty caused by the replacement of the military by a civilian agency to manage the manned space programme, resources for Mir would bring the completion of the station to a standstill.

2.2 RUSSIAN ACCOMPLISHMENTS ON MIR

Originally, Mir was to have been fully assembled in 1990, complete its five-year design lifetime soon afterwards, and then be replaced with Mir 2 in 1992. As funds for the civilian space programme diminished, the Russian political leadership and the Russian Space Agency realised that there would be no Mir 2. The final two laboratory modules for Mir, which would be completed with assistance from the US, were finally completed and launched to Mir after the Administration of President Bill Clinton forged an agreement with Russia for a joint manned programme in space.

In a paper published in the *Journal of the British Interplanetary Society* in 1997, Andy Salmon carried out an extensive survey of the science experiments conducted on the space station between 1986 and 1994, before the beginning of US missions to Mir, and before the last two modules were added.

Experiments were conducted, with varying degrees of success, in protein crystal growth, human physiology and psychology, materials science, radiation in the space environment, Earth observation, geophysics, bioprocessing, biology, astronomy and new technologies. Much of the equipment and instrumentation was provided by international partners.

One of the most interesting and successful series of studies performed on Mir, which began before cooperation with the US and continued throughout the Shuttle–Mir programme, was the study of plants in greenhouses designed for space. Just after the turn of the twentieth century, Russian space pioneer Konstantin Tsiolkovsky had described such space gardens in his imaginative while technically detailed visions of living in space.

Experiments to grow plants in space – needed eventually both to supply nutrients for an interplanetary crew and to also produce oxygen for their life support – were begun on early Soviet space stations, with limited success. It became clear that the process of providing the delicate conditions for plants to flourish – proper light, controlled temperature, appropriate levels of carbon dioxide, a workable water delivery system – was more difficult to orchestrate than had been expected.

Early 'hothouses' that were brought up to Salyut 6 failed to bring plants to the flowering stage. A series of plant growth units were used on Salyut 7, and plants were grown in the base block of Mir by the first crew to live there. The series of crop

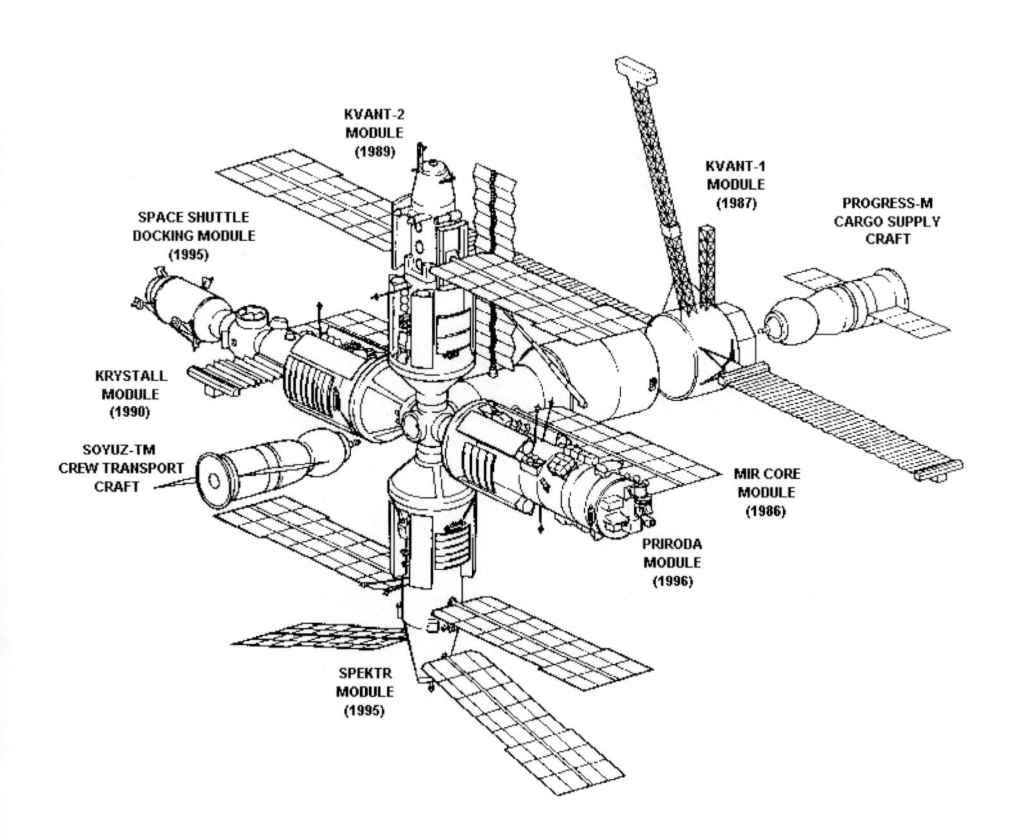

Figure 2.2. The Mir space station at 'assembly complete' in 1996. (Courtesy Energia.)

growth experiments in the Svet ('Light') greenhouse, provided by the Bulgarian Academy of Sciences, will be described in detail in the next chapter.

One of the functions of growing plants in space that the Russians have investigated is the ability to reduce the psychological stress of spaceflight. The Vozon greenhouse on Mir, from the Botanic Garden of the Academy of Sciences, used the growth of orchids and a dwarf magnolia vine for such a purpose.

In inorganic materials sciences, Russian and international investigators, notably from Western Europe, have conducted experiments in a series of furnaces onboard Mir, examining new materials for semiconductors, superconductivity applications, and metal alloys. Salmon reports that many of the results were only partial because of power shortages on Mir (the furnaces are very electricity-intensive), variations in the level of microgravity on the station, equipment failures and breakdowns, and the inability to return a large number of samples to Earth for intensive examination. A Soyuz spacecraft, returning two or three crew-members to Earth, can carry 50–120 kg of cargo.

To help alleviate these constraints, in 1990 the Soviet Union developed the

Figure 2.3. In order to allow for the return to Earth of material from Mir, the Russians developed the Raduga Recoverable Ballistic Capsule, seen here before a flight on an unmanned Progress supply ship. (Courtesy Mark Wade.)

unmanned Raduga Recoverable Ballistic Capsule, specifically for the return of high-value science payloads. In a presentation to a two-day symposium, 'Space Station Mir: A Technical Overview,' held on 27–28 July 1993 in Virginia, sponsored by Mir's manufacturer and manager, NPO Energia, Viktor Minenko described the development and capabilities of the Raduga.

The payload compartment of the Raduga is designed to return to Earth film and magnetic tapes with data, biological materials and samples from biotechnology experiments, samples of new materials developed in the furnaces onboard Mir, and some equipment and samples to study effects such as the radiation environment and deterioration of materials outside the station.

Raduga is also, as Minenko described, 'an attractive means for conducting physical and technical research during the phase of descent' through the Earth's atmosphere, including the heating of materials as the capsule speedily descends.

Observing the Earth from space has been a favourite pastime of all space travellers since the start of the Space Age. The Mir crews had an array of Earth-observing cameras and instruments to image the planet in black-and-white, colour, infrared, and a range of frequencies in the electromagnetic spectrum to measure ground and atmospheric characteristics.

Observing the heavens was also a major activity onboard Mir. As described above, the Kvant 1 laboratory focused on astronomical observations, and was able to be controlled from the ground. Foreign university science teams also participated, and the crew interacted with the instruments only if there were errors in commands sent to the onboard computer, or if it were necessary to make adjustments.

Many of the limitations that scientists and cosmonauts were faced with in performing experiments on Mir were improved with participation from the United States, with its sophisticated electronic and communications systems, and a Space Shuttle that on each flight can carry tons of equipment, samples and data back to Earth.

2.3 EXPERIMENTS IN THE LIFE SCIENCES

According to O.V. Mitichkin, speaking at the July 1993 Energia symposium, during the eleven years from 1982 to 1993 more than 100 biotechnology experiments were conducted on the Salyut 7 and Mir stations. To carry them out, 14 types of specialised unit were employed. The first research conducted was the electrophoresis of proteins and cells. The treatment of various diseases depends upon the accessibility of very pure biological materials that are difficult to separate in the gravity of Earth. These include specialised cells, such as those that produce insulin, or proteins that can play a role in combating disease.

Electrophoresis is the process of separating different materials in a solution through the application of an electrical current. In Earth-based laboratories, the fact that particles of different size move at different rates under the influence of an electrical field has been used to separate the components of natural fluids such as blood sera and protein solutions, to be able to analyse them. But under the influence of gravity, convection currents created by the heat produced from the electrical field introduce turbulence, which interferes with the process of separation.

Mitichkin stated that about 70 experiments to separate various proteins and cells had been conducted up until 1993 in electrophoresis equipment, using four different methods. The results, he reported, verified the theoretical hypotheses on the increase of purity and productivity of the electrophoretic method. Some of the most difficult materials to separate on Earth – organic cells and antibiotic products of the spores of microorganisms – were separated in space.

There were instances where the productivity of separation increased from 300 to 1,000 times better than on Earth, and the degree of purification increased up to ten times. One of the materials that was processed was alfa-interferon, which was produced in space free from inactive forms of interferon that can cause harmful side effects on the human immune system. Electrophoresis experiments also involved separation of an anti-influenza preparation, later used as a standard for a vaccine, the production of antibiotics from microorganisms that were of high quality.

Another series of experiments involved the growth of bacteria in microgravity, where some important differences were observed from those on Earth. Mitichkin reported that experiments in the Recomb facility on Mir, involving the growth of

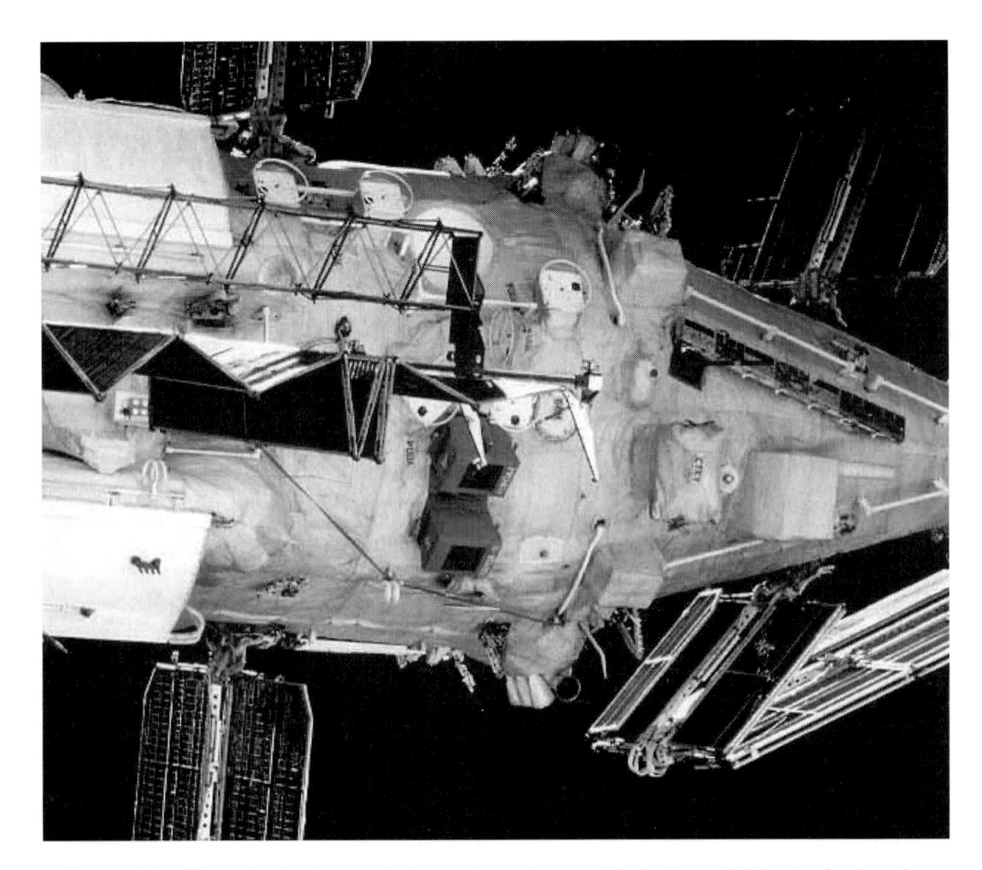

Figure 2.4. When the Spektr module was launched to Mir in June 1995, attached to the outside were the two Kozma instruments, designed by scientists from the Space Research Institute of Russia. These interstellar-atom collectors are seen in the centre of the photograph taken by the *Atlantis* Space Shuttle that autumn, appearing as two protruding boxes. Samples of foil were exposed to the interstellar medium, and returned to Earth for examination.

plant tissue and bacteria, indicated that new hybrid cells which are not stable or cannot exist on Earth, were achieved in space.

Experiments with various strains of genetically altered *E. coli* bacteria dramatically increased the transfer of genetic material from a cell donor to a recipient. Since such genetically altered strains can be used to produce biologically active substances such as amino acids, hormones, vitamins, and antibiotics, this could have significant practical value. The efficiency of the transfer of complete bacterium chromosomes increased up to 74% on Mir compared to 16% on Earth.

The Russians and their international partners carried out experiments in cell culturing using the MKM-1 (vita) unit, built by the Crimean Medical Institute in Ukraine, the Mendeleev Moscow Chemical Technological Institute, and the Biopreparat Corporation, under the Ministry of Medical Industry, in order to

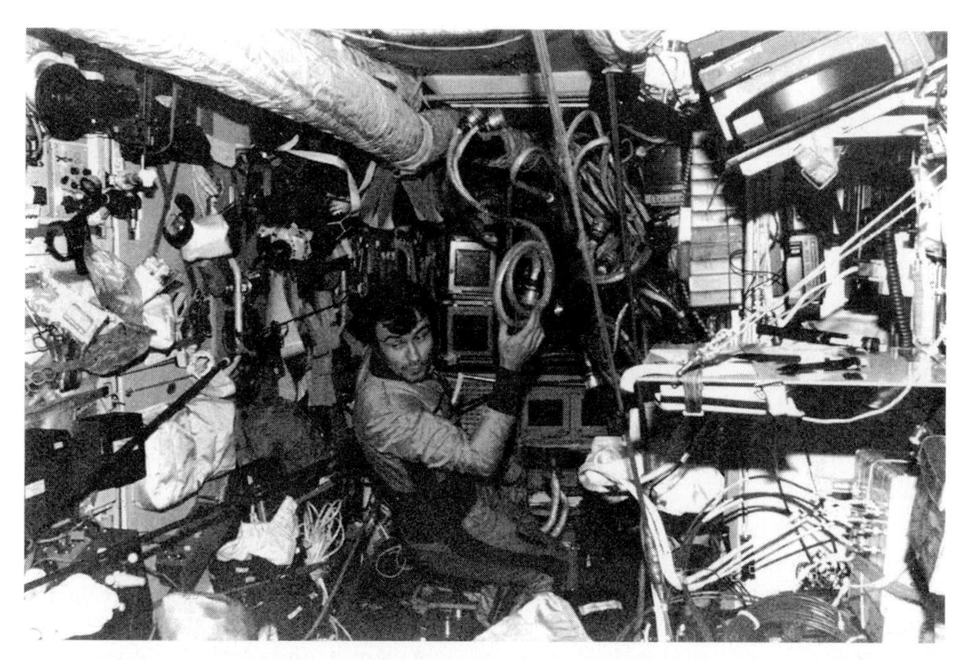

Figure 2.5. Mir 18 cosmonaut Vladimir Dezhurov reroutes cables as part of the activation process in the Spektr. Two years later the module would have to be abandoned after an errant Progress supply ship punctured the laboratory, causing it to depressurise. To prevent the atmosphere from the entire station from escaping into space, the cosmonauts had to quickly sever the cable connections between the Spektr and the rest of the station.

study the basic processes of cell growth, to perform the cultivation of genetically engineered microorganisms, and to cultivate proteins in the membranes of cells, among other applications.

From 1991 to 1993, experiments were conducted on the cultivation of tissues and cells of wheat and potatoes, during the Maksat experiment. Results indicated that under the powerful impact of the stress of microgravity on plants, there was more intensive growth, and an increase is their vitality and immunity to temperature variation, toxin, viruses and other stresses, when tissues grown in space were replanted on fresh nutrient media on Earth. Mitichkin also reported that through the Oner experiment to grow fungus products on Mir, there was a development of clones with a greater ability to produce antibiotics in cultures delivered from space.

Protein crystal growth experiments began on Mir at the end of 1987, when the Azerbaijani Ainur facility was delivered to the station. By 1993 more than ten types of protein had been grown in the biocrystallisation facilities on Mir. Using both liquid and vapour diffusion techniques, samples up to 0.4–2 mm in size were grown, which were larger than those grown on Earth, and with better structural characteristics. Crystals included those of genetically engineered human growth hormone, the influenza virus, and other proteins important for human health.

The American company Payload Systems Inc flew protein crystal experiments twice on Mir, in 1989 and 1992. In the later experiment, proteins provided by Canadian, German, Japanese, Dutch and American researchers were flown. And while a November 1992 article in *Nature* contained complaints that an improvement in *only* 24% of the crystals flown was achieved, this was in fact a significant improvement over the 20% achieved on shorter Space Shuttle flights.

Although the Soviets, in 1993, hoped that continued research on next-generation and more sophisticated and automated biotechnology facilities would lead to the commercial development of various products, the expansion of scientific capabilities onboard Mir would, in general, take place when the Americans became part of now-Russian manned space activities.

2.4 ASSEMBLY COMPLETE

The Spektr module was launched to Mir on 20 May 1995, while the first American astronaut Dr Norm Thagard was onboard the space station. Spektr was to perform multiple functions. It had four solar arrays attached, in order to provide an additional 16 kW of electric power. Originally, according to David Harland, Spektr was designed to be an Earth-observation laboratory, and carried a variety of multispectral sensors for that purpose.

But in addition to its function as a power and remote sensing facility, Spektr carried 880 kg of American equipment for Thagard, mainly to enable more sophisticated observations of the human physiological adaptation to microgravity. This new module also became available space as research and living quarters for US astronauts. The module became famous on 25 June 1997, when an unmanned Progress supply ship collided with Spektr, resulting in the first depressurisation of a spacecraft on orbit.

The final laboratory for Mir, called Priroda (Nature), was launched on 23 April 1996. The focus of studies on Priroda was the study of the Earth's environment, including sensitive measurements of the thermal variations, energy levels, and the height of waves of the world's oceans. Other studies included taking measurements of the propagation of pollutants, temperature and vertical profile of clouds, wind direction, and other characteristics of the Earth's atmosphere.

Priroda was also packed with 650 kg of scientific equipment, and other supplies for astronaut Dr Shannon Lucid, who had arrived on Mir a month earlier. This included facilities for biomedical studies and also microgravity experiments. One piece of equipment Lucid assembled – which would remain on Mir and do a yeoman's job throughout the entire joint Shuttle–Mir programme – was the Space Acceleration Measurement System (SAMS), which recorded vibrations onboard the station. The Canadian-built Microgravity Isolation Mount was a complementary unit, intended to isolate microgravity experiments and eliminate the vibrations caused simply from the crew's movements and activities on the station.

With the integration of Priroda into the Mir complex, the space station was complete. For the next three years, through trials, tribulations, accidents and crises,

Figure 2.6. Respected scientist Academician E.P. Velikhov warned Russian and American policymakers through the early 1990s not to allow the destruction of science following the collapse of the Soviet Union. Here he is speaking at a press conference in Washington DC in 1990. (Courtesy Stuart Lewis/EIRNS.)

the station would nonetheless prove its worth as a platform for scientific study in space.

2.5 THE SHUTTLE–MIR PARTNERSHIP

As the political upheaval in Russia and the dissolution of the Soviet Union were underway, a debate was being carried out in Washington as to how to orient to this new political geometry in eastern Europe. Prominent Russian scientists were issuing calls for help.

Academician E.P. Velikhov, Vice President of the Soviet Academy of Sciences, prominently involved in all major Soviet science programmes from thermonuclear fusion to laser defense, presented a speech at the Extraordinary USSR Congress of People's Deputies on 2 September 1991, pleading with the deputies not to break apart the scientific establishment. 'What distinguishes Third World countries from those in the First World?' he asked. 'In the main, Third World countries have resources, they have a work force, too, but they do not have science or expertise... If we destroy science, we shall never rebuild it... then we will have no future... Science is a very delicate instrument, and it is now collapsing very quickly.'

Dr Roald Sagdeev, former Director of the prestigious Soviet Space Research Institute in Moscow, and currently Professor of Physics at the University of

Figure 2.7. Professor Roald Sagdeev has been a strong proponent of US–Russian collaboration, and warned that the Soviet space capabilities, which are 'international treasures', were threatened with extinction due to the decline of the Russian economy. Here Dr Sagdeev engages in conversation with Senator Barbara Mikulski at a space symposium in 1992. (Marsha Freeman/*21st Century Science & Technology*.)

Maryland, echoed Velikhov's concerns at a hearing before the House Committee on Science and Technology on 21 February 1992. 'The collapse of the Soviet Union has left its space assets an endangered species. In the current economic climate it seems highly unlikely that Russia, or any other independent state of the newly born Commonwealth, could carry a space programme even remotely similar in size and quality to that which has developed since 1957... The assets,' he stated, 'represent quite an elaborate and broad network of design bureaux and enterprises, of space industry, and scientific institutions for space exploration.'

Most important, Dr Sagdeev told the American congressmen, was that 'Soviet space assets could be considered as truly international treasures, and it would be extremely painful if, in the process of economic disorder and disintegration... these tremendous achievements of humankind would be lost.'

Congressional representatives were pressuring the Bush Administration to enter into partnerships with the former Soviet scientific community, if for no other reason than to take advantage of the unique technology that they were now willing to share with the US. But representatives of the American Defense Department and the Department of State were hemming and hawing about the 'security risk' of sharing technology, objecting that any aid to Russia's former military/industrial complex, including the civilian space programme, would be a security threat to the US. Their view was that allowing the former Soviet Union to become weaker would somehow make the United States stronger.

Opening a hearing on 25 March 1992, in the throes of the Russian policy debate, House Science and Technology Committee Chairman George Brown, from California, stated: 'Today there is universal recognition that a strong science and technology base is fundamental to the strength of an industrial economy... The Soviet Union once maintained the largest scientific establishment in the world. Fully one quarter of the 'scientific workers' on the planet now reside in the republics of the former Soviet Union, and more than half the world's engineers work there.' Representative Brown reported that in his discussions with Russian scientists, they stressed that even modest cooperative efforts between Russia and the US 'could maintain research efforts of potential significance to the entire world.'

The hearing included a teleconference with the leadership of the Russian scientific community in Moscow. Academician Velikhov stressed that the disarming of the former Soviet military establishment had left factories and enterprises without the resources needed to convert to civilian products, as the Russian government had mandated. If the US would help defence factories to retool, he said, it would not have to worry about increased military production.

Dr Edward Teller, who has worked on advanced nuclear weapons projects in America since World War II, stressed that the best way to know what the Russian military was doing was to have joint programmes, not by trying to protect US 'secrets'. He was joined in this evaluation by the former ambassador to Moscow, Jack Matlock, who warned that the policy choice facing the United States was either to take the post-World War I Versailles Treaty path to cripple the economy, which led to World War II, or the post-World War II Marshall Plan approach of rebuilding, which created democratic nations, and allies, from former enemies.

In 1988, President Ronald Reagan and Soviet President Mikhail Gorbachev agreed that scientific instruments would be launched on each other's spacecraft. In July 1991, at a Bush–Gorbachev summit, discussions were held on upgrading space cooperation. In June the following year, at a summit between Presidents Bush and Boris Yeltsin, an agreement was signed to allow both nations' space agencies to pursue a broad range of possible joint projects. On 18 June, NASA and the Russian Space Agency signed a $1 million contract for the US to begin to evaluate the use of Russian technology for Space Station Freedom.

The Shuttle–Mir programme was initiated after a trip to Russia by NASA Administrator Dan Goldin later that summer. The plan called for a Russian cosmonaut to fly on the Space Shuttle, and for an American astronaut to visit Mir for three months. The primary reason given for the programme was to extend the time during which US life sciences research could be conducted in space, which was limited to a maximum of two weeks on the Space Shuttle.

In October 1992, US and Russian space officials signed an agreement to fly a cosmonaut on the Space Shuttle in 1993, and an astronaut to fly with two cosmonauts on a Soyuz to Mir, and spend about three months there, in 1995.

In April 1993 the new President, Bill Clinton, held his first summit with Russian President Yeltsin in Vancouver, Canada. This resulted in an agreement to expand the cooperation beyond the 1992 agreements, as part of the President's effort to strengthen America's relationship with Russia. Soon afterwards the American President requested NASA to restructure the Space Station Freedom programme, and in September the Russians were invited to participate in that activity so that their potential contribution could be examined. The Russians had to finally admit that they did not have the resources to build their follow-on Mir 2 station.

In November 1993 a three-phase plan was detailed for cooperation between the United States and Russia. There would be up to ten flights of the Space Shuttle to an upgraded Mir station, with up to two years of astronaut time on Mir, and a joint science programme. The Shuttle–Mir programme became Phase I of a three-phase programme, to culminate in the building of the International Space Station (ISS), with the European Space Agency, Japan, and Canada.

At the close of 1993 it was agreed that a $400 million contract would be given to the Russian Space Agency to pay for American use of Mir, and for hardware to be purchased from Russia to carry out the joint programme.

On 24 June 1994, NASA Administrator Dan Goldin and Russian Space Agency Director Yuri Koptev signed a contract under which NASA committed itself to the 'enhancement of Mir 1 operational capabilities; joint space flights; and joint activities leading to Russian participation in the design, development, operation, and utilisation of an International Space Station.'

But by 1996 it was clear that the Russian Space Agency did not have the resources to keep the bargain and meet its commitments to deliver hardware to the ISS on schedule. The US–Russian agreement was extended again and modified. Russia agreed that it would deliver two modules on time for the ISS, and would build a cargo spacecraft called the Logistics Transfer Vehicle. NASA agreed to pay Russia $72 million more than the original $400 million, and exercised the option for two

Figure 2.8. Yuri Koptev, General Director of the Russian Space Agency, has led the Russian effort to merge the world's only two manned space programmes.

more Shuttle–Mir docking flights to help Russia keep Mir operational longer than originally planned. The Shuttle flights to Mir would deliver supplies that would otherwise have to be brought to the station by Progress supply ships.

As Frank Culbertson, NASA astronaut and manager of the Shuttle–Mir programme, explained in testimony to the House Committee on Science on 18 September 1996: 'Initially, the role of the US crew was patterned after that of the other foreign personnel, to fly to the Mir as guest cosmonauts.' But the United States was now entering into a partnership with the Russians. 'As the Shuttle–Mir missions progressed, it became clear that our goal of learning how to work with the Russians should include direct knowledge of the operational techniques, through involvement in the operations themselves,' Culbertson explained. 'The Russians quickly agreed to the principle of making our astronauts an integral part of the crew, and work was begun to modify the training programme to allow for expanded duties, with some changes even made before Shannon Lucid's mission, including the change from Cosmonaut Researcher to Flight Engineer-2.'

As the cooperation changed to collaboration, the goals expanded from carrying out biology and life sciences experiments to learning how to operate a space station. According to Culbertson, the goals of Phase I (the Shuttle–Mir flights) were:

1. To learn to work together, both in space and in ground support activities.
2. To reduce the risks to ISS development and operations by testing hardware, refining joint procedures, and integrating the operational practices of the two nations which have primary operational responsibility for the ISS.
3. To gain experience in long-duration stays on a space station, and develop effective biomedical countermeasures to the effect of extended weightlessness.
4. To conduct scientific and technological research in a long-duration environment, gaining valuable research data, and developing effective research procedures and equipment for use in the ISS.

2.6 AN ORBITAL BALLET

One of the most important and most basic accomplishments of the Shuttle–Mir programme has been to learn to rendezvous and dock the Shuttle to Mir. The Shuttle had never before docked to any manned spacecraft.

Although there was a link-up between an Apollo command module and a Soviet Soyuz spacecraft almost 20 years before the first Shuttle–Mir docking, that singular event never developed into a long-lived space collaboration. But it was a pathfinder for the joint programme of the 1990s.

The hope that following Apollo–Soyuz, joint docking flights would continue – at least to provide rescue capability should either nation's astronauts need assistance in space – was dashed with the Soviet invasion of Afghanistan in 1979, and the imposition of martial law in Poland in 1981. But the unique capabilities independently developed by each manned space programme during the 20-year hiatus in joint programmes are complementary to each other today.

Figure 2.9. Space Shuttle STS-71 Mission Commander Hoot Gibson described the intricate manoeuvres required to rendezvous and dock the orbiter *Atlantis* with Mir as an 'orbital ballet'. The sheer size and mass of the two vehicles presented a challenge. This photograph, taken by the Mir 18 crew on 4 July 1995, shows the two behemoths at the end of the first docking mission of the Shuttle and Mir.

Bringing together two spacecraft as massive as the Shuttle orbiter and Mir required the most delicate piloting skills. As soon as the Space Shuttle *Atlantis* lifted off at 3:32 pm on 27 June 1995, it started on its approach to the Mir station. While the Mir 18 crew onboard the station reported that they were 'tidying up' for the first visit of the Shuttle, Commander Hoot Gibson was bringing *Atlantis* within range of Mir.

Unlike the chase by the Shuttle to catch up with the Hubble Space Telescope or other satellites with which astronauts have rendezvoused, the Shuttle took advantage of orbital mechanics, raising its altitude to slow itself in order to minimise the firing of its thrusters in proximity to the station. *Atlantis* approached Mir from underneath in order to avoid thruster firings which could have damaged the station's delicate solar arrays.

The docking of these two massive vehicles was described by Shuttle pilot Charles Precourt, during a pre-flight press conference, as an 'eight-dimensional problem'. The crew had to control the orbiter's own axes of rotation to keep its attitude steady in the pitch, roll and yaw directions. The second set of three dimensions concerned the Shuttle's position relative to Mir – up/down, left/right, and forward/aft. In order to soft dock the two vehicles, the Shuttle had to make the final approach to Mir at a 'glacial' speed of between 0.02 and 0.04 metres per second.

The eighth constraint was time. The Russian Mission Control Center in Kaliningrad had to be able to communicate with the cosmonauts onboard Mir during the docking sequence to allow real-time back-up commands if needed, and to be able to coordinate the docking with Mission Control in Houston. The Russians can communicate with their cosmonauts only when Mir is over one of the Russian ground stations, which, for security reasons, had been built only in the Soviet Union.

During the Shuttle–Mir mission, Russian Mission Control was able to communicate with the crew for 40 minutes out of each approximately 90-minute orbit, for the first nine hours of their day, and hardly at all for the next nine hours. This translated into a 2–3-minute envelope within which Commander Gibson had to perform the sequence of docking manoeuvres on the Shuttle. The United States, by contrast, uses a constellation of six geosynchronous satellites – the Tracking and Data Relay Satellites – which provide the capability to transmit voice, data, tracking information and telemetry between the ground and the Shuttle for 85% of the Shuttle's orbit.

During the five days that the two spacecraft were docked, engineers on the ground had the opportunity to compare the predictions from their computer models of how the entire stack would behave, to the real thing. After the first day of docked operations, NASA flight controllers reported that the propulsion fuel that the Shuttle was expending to hold the entire stack of Mir and the Shuttle in the attitude needed to point Mir's solar arrays toward the Sun to produce electricity, was 70% higher than projected.

Before the flight, engineers had determined the energy requirements for attitude control using models based on estimates of total mass, mass distribution, stiffness of the structure, and other variables which could be determined accurately only when the two spacecraft were docked. Relatively minor adjustments, on both the Shuttle and Mir, which were coordinated closely between the two Mission Control Centres, reduced the fuel use.

There were many 'firsts' on this joint mission. One of the most striking was the sight of three bodies flying separately, but in close proximity, as the undocking sequence unfolded after *Atlantis* had completed its mission to Mir. Commander Gibson had said that during the training for this in the simulator, 'the words 'cosmic ballet' came to mind.'

Early on 4 July, the Mir 19 crew – Commander Anatoly Solovyov and Flight Engineer Nikolai Budarin – left Mir and entered their Soyuz capsule. At 7 am the small Soyuz – which brought the cosmonauts up to the Mir, remained attached to it during the mission, and then returned the crew to Earth – undocked from the top of the Mir. The Shuttle, still attached to the bottom of Mir at the Kristall module, took photographs as the Soyuz slowly 'floated' away from the huge complex.

At a distance of about 100 metres from Mir, the Soyuz stopped and photographed *Atlantis* as it undocked from the station. When *Atlantis* was about 230 metres away from Mir, it held that position. After the Soyuz had redocked to Mir, the Shuttle began a 360-degree fly-around of the station, to provide the Russians with photographic documentation of the overall state of Mir after nine years in orbit.

In important ways, the flawless docking mission followed in the footsteps of

Figure 2.10. The only precedent for the joint Shuttle–Mir flights was the 'handshake in space' during the Apollo–Soyuz mission in 1975. The trust and friendship that developed between the commanders of those two spacecraft, General Alexei Leonov (left) and General Tom Stafford, was a long-lasting result. Here they are seen at the Baikonur launch facility in Kazakhstan, during the Apollo crew's training in Russia.

Apollo–Soyuz. Many of the engineers on both sides who worked on the 1975 mission, worked on this mission. The docking mechanism that was used for the Shuttle and Mir was modelled upon the mechanism that first joined the US and Soviet space programmes two decades earlier. Most important, a lasting trust and friendship developed, especially between the two commanders, General Tom Stafford and General Alexei Leonov, who provided guidance for the Shuttle–Mir project.

While preparing for the Apollo–Soyuz mission, the US astronaut crew – Tom Stafford, Donald 'Deke' Slayton and Vance Brand – visited the Soviet Union only twice during more than two years. Each crew was launched on its own spacecraft, met briefly in space, and returned home. In comparison, during the *Atlantis*–Mir mission, two cosmonauts (the Mir 19 crew), and three astronauts were flown up on *Atlantis*, and two different cosmonauts (the Mir 18 crew), who had been launched on a Soyuz in March, came back on *Atlantis*.

Dr Norm Thagard became the first American astronaut to fly on a foreign vehicle when he was launched on a Soyuz with his Mir 18 colleagues in March. Therefore, a total of four cosmonauts and one astronaut (or half the ten-person combined

Atlantis, Mir 18 and Mir 19 crews) had flown on *both* American and Russian spacecraft when the Mir 19 crew landed in September. This required the closest working relationship between the crews, who trained together in both Star City in Russia and at the Johnson Space Center in Houston.

2.7 IT RUNS LIKE A RAILROAD

For the Americans, one of the most impressive features of the Russian space programme was the efficiency with which the Mir space station was serviced in orbit. Although the United States has docked vehicles in space, notably during the lunar-orbit rendezvous of the Apollo missions to the Moon, there is no question that the Russians have become the experts at docking and undocking different kinds of spacecraft in orbit.

It was necessary for the Russians to develop this capability for their manned programmes because they do not have any vehicle near the size of the Shuttle for transporting crews and supplies to orbit. Therefore, if a crew was to stay on Mir for months at a time, supplies had to be delivered periodically. It was also useful to have an unmanned Progress cargo vehicle 'visit' Mir during long-duration flights, to take additional scientific equipment plus letters from home and treats for the cosmonauts. The Progress arrived at the station, docked there automatically, and was jettisoned to be burned up in the atmosphere after its cargo had been unloaded and traded for trash from the station.

The confidence which the Russians have developed in their ability to move modules around to reconfigure the station when necessary, was demonstrated in the hectic activity in preparing for the *Atlantis* docking during the STS-71 Space Shuttle mission, to ensure that the orbiter would not damage Mir.

In the space of six weeks, the Russian cosmonauts of the Mir 18 crew performed five EVAs, moved the Kvant 1, Kristall, and Spektr laboratory modules to different ports, which required the repositioning of their one docking cone each time, moved solar arrays between modules, and attempted to repair the balky solar arrays. As astronaut Norm Thagard said during a press conference onboard Mir on 12 April: 'The Progress comes up on time, the Soyuz comes up on time. It runs like a railroad, and when you look at the budget its operated on, that's pretty impressive.'

Each national space programme has its own strengths, and confidence in its own systems and technology. The Russians have accumulated a great deal of experience with orbital structures, which was evident as soon as there was a problem during the STS-71 mission. At the time that the Shuttle undocked from Mir, the space station lost attitude control and started to rotate out of position. Mission Control in Kaliningrad ordered the crew in the Soyuz spacecraft back to redock with the empty Mir five minutes earlier than planned, before its misalignment would present a problem in redocking the Soyuz. When nervous American engineers at Mission Control in Houston expressed their concern, Russian flight controllers assured them that this had happened before on Mir, and they knew how to fix it (which they did).

The shoe was on the other foot during the earlier STS-63 Shuttle rendezvous

Figure 2.11. Keeping Mir operational and preparing for the arrival of new modules or the Space Shuttle entailed a dizzying schedule of space-walks by the cosmonauts. In this picture, Mir 18 flight engineer Gennady Strekalov is seen during one of five such space-walks during his increment on Mir.

mission in February 1995, when a leaky thruster on *Discovery* threatened to ruin a planned first close rendezvous between the orbiter and Mir. NASA analysed the problem, and was convinced that the leak would not damage the Mir, and that Commander James Wetherbee could manoeuvre the Shuttle even if one leaky thruster were shut off. The nervous Russians said they would give the go-ahead for a rendezvous, but only to a distance of 130 metres. Throughout the night, engineers halfway around the world conferred, and on 6 February Wetherbee received the go-ahead and brought *Discovery* within the planned 13 metres of the space station.

2.8 BIOMEDICAL RESULTS FROM MIR

Although the Russians have operated manned space stations since the early 1970s, the data they have collected on the effects of the microgravity environment on the cosmonauts have been poor or non-existent.

The Russians have not developed the kind of highly sophisticated biomedical equipment that is used in the Space Shuttle programme. Due to a lack of both the physical space for more hardware, and of the availability of compact, high-quality equipment, the Russians have carried out almost no in-flight analysis of physiological changes – just pre- and post-flight studies.

On the very first Shuttle–Mir mission, the European Spacelab enclosed laboratory module was housed in the orbiter's payload bay, and was outfitted for a full set of life sciences experiments. The Mir 18 crew underwent extensive testing of body functions that are known to change in microgravity, in the cardiovascular system, the skeletal and muscle systems, and the immune system. Spacelab carried, for example, a portable blood analyser to check the astronauts' blood chemistry while they were still in orbit, without waiting for samples to be returned to Earth.

Space Shuttle Mission Specialists Ellen Baker (a medical doctor) and Bonnie Dunbar (a biomedical engineer) carried out invasive experiments, such as injecting the crew with pneumococcus to perturb their immune systems so that they could see how they would respond. Dr Helen Lane, the principal scientific investigator for metabolic studies, explained at a briefing after the completion of the mission that an advantage that the United States has is that the landing sites for the Shuttle (either California or Florida) are not as remote as the Russian landing site (Kazakhstan), allowing the biological samples to be received in laboratories in only a matter of hours. The long-duration Mir 18 crew-members also underwent *in situ* tests to study how much bone calcium had been lost in weightlessness.

One of the most severe constraints on the Mir–Soyuz system for scientific research is the lack of room to bring back material to Earth. Unlike the Shuttle, which has a

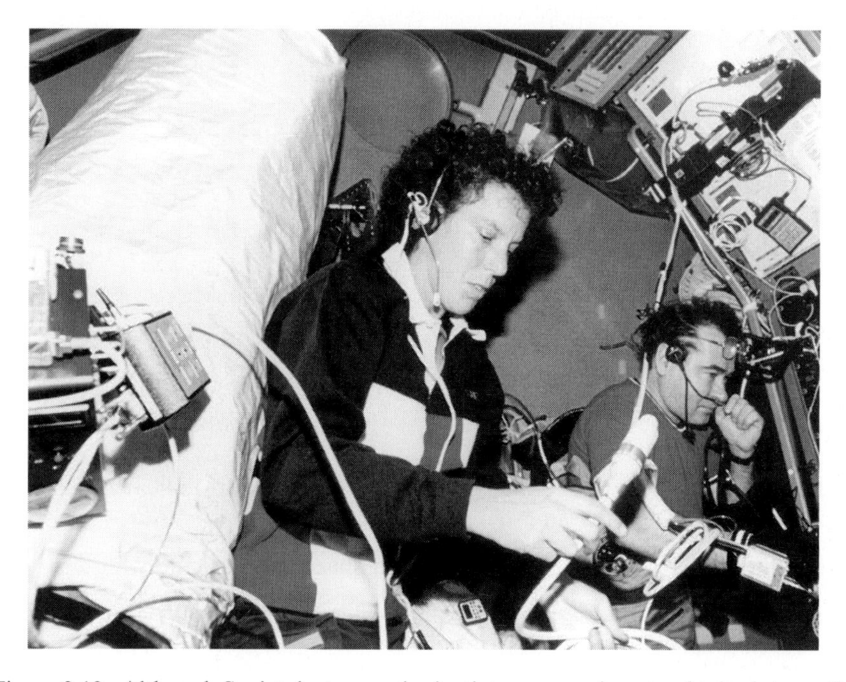

Figure 2.12. Although Soviet doctors and scientists were anxious to obtain data on the human body's adaptation to the weightless of space, and attempts were made to do so on all space station missions, the addition of more sophisticated medical technology through the Shuttle–Mir programme greatly improved the information available. During the first Shuttle–Mir mission, astronaut and medical doctor Ellen Baker monitors cosmonaut Gennady Strekalov during a run on the Spacelab treadmill in the Shuttle.

maximum 60,000-pound payload capability, the small Soyuz capsule can typically bring home less than 100 kilograms of payload with two cosmonauts on board. This has severely limited the biological samples of, for example, urine, blood and saliva that the crew can transport to Earth for analysis in laboratories. For the joint missions, the Shuttle carried hardware up to Mir, and returned important biological samples and other materials back to Earth.

Although the crammed schedule of tests and experiments in Spacelab during the first docking mission – especially during the five days that the Shuttle and Mir were docked – sometimes tested the patience of the astronauts, after the Shuttle landed, mission scientist Tom Sullivan reported that they had accomplished 10% more scientific work than had been planned. A week later NASA life sciences Director Dr Arnauld Nicogossian reported that based on these preliminary findings, there did not seem to be any 'show stoppers' for long-duration spaceflight.

The seven long-duration missions with US astronauts produced a treasure trove of experimental data on the human adaptation to weightlessness, and provided a test bed for new health care technologies that will be used on the International Space Station. In a monograph published after the completion of the Phase I Shuttle–Mir programme, NASA summarised the conduct of research on Mir that has an impact on human health.

One concern for long-duration space missions, but not often addressed, is the ability to treat unforeseen medical events onboard a spacecraft. During the Phase I programme a cardiac defribrillator was delivered to Mir to upgrade the station's emergency medical treatment capabilities. The technology for this portable device has been developed from NASA research, and is now also used on commercial airline flights.

In order for a crew in space to have access to medical expertise on the ground, NASA is working with an international group of physicians to develop advanced telemedicine systems. During the Shuttle–Mir programme, astronauts and cosmonauts were in contact with medical doctors on the ground through a visual and audio communication system, with the intent to pioneer use of the Internet to put the most advanced medical knowledge available anywhere in the world in the astronaut's hands.

Examination of the data on crew health returned from Mir indicated that there is an 'apparent increase in cardiac dysrhythmia after about 70 days of spaceflight'. Medical specialists believe this is a benign response to weightlessness, but the phenomenon is still being studied.

Observations and measurements from Skylab on the loss of bone mass by astronauts over the entire time of their stay in microgravity was confirmed from the Mir crew data. It was observed that while adaptation to space is sometimes described as accelerated ageing, the rate of bone mass loss is three to ten times *greater* than the loss associated with ageing on Earth. Phase I revealed that exercise is only partially effective in reducing the loss of muscle mass, which, like bone loss, showed a continual decline.

Adaptation to microgravity causes a shift of bodily fluid up toward the upper torso and head, as there is no gravity to pull it down to pool in the lower extremities.

The body, sensing that there is then too much fluid, increases urinary output to eliminate the perceived excess. At the same time, the calcium due to bone loss is being excreted, leading medical experts to be concerned about an increased risk of developing kidney stones in space.

While there has so far been no report of an astronaut or cosmonaut developing such a problem while in flight, a comparison of the data from 150 Space Shuttle astronauts with Phase I astronauts and cosmonauts has suggested that the risk of developing kidney stones is directly proportional to the time spent in space. To date, only three astronauts have developed kidney stones – two preflight and one post-flight.

Research on Mir revealed that longer exposure to the space environment leads to more significant changes in neurosensory functions, which include those that involve coordination between the brain and the senses. These changes include less control over posture, the deterioration of eye–head coordination while moving, and difficulty in fixing the eyes on a moving target.

2.9 THE CRISES ONBOARD MIR

On 23 February 1997, while American Dr Jerry Linenger was onboard Mir as the third long-duration astronaut, a fire broke out as the crew was activating an oxygen-producing canister. There are discrepancies among the crew-members' descriptions of the length and extent of the fire, but it was certainly a most serious accident to occur on a spacecraft.

Figure 2.13. Dr Jerry Linenger was the third NASA astronaut to live onboard Mir, replacing John Blaha. In this photograph he is shown soon after assuming his duties on the station in January 1997. A month later a fire on Mir would lead him to state that the station was not safe, and that no other astronauts should be sent.

The events surrounding the fire on Mir have been described in great detail by Bryan Burrough in his 1998 book, *Dragonfly: NASA and the Crisis Aboard Mir*. One result of the fire was that the event apparently traumatized Linenger. On 16 May, the Space Shuttle *Atlantis*, commanded by Charles Precourt, docked with Mir to return Linenger to Earth and deliver astronaut Michael Foale to the station. Linenger said that he opposed Foale's staying on Mir, saying that he did not think it was safe.

Shuttle Commander Precourt had been asked by NASA managers to assess Linenger's state of mind after *Atlantis* docked with Mir. '[Linenger] sounds like what he is, a rookie,' Precourt reported. Later, he told Bryan Burrough: 'You know, I've been around fighter jets and spacecraft for 25 years, and these things just happen. You have a fire in your engine and you just don't quit. You fix it and go on. You just go on.'

Mir's troubles were far from over. Four months later, on 25 June, an unmanned Progress supply ship collided with the Spektr module of the Mir station, as commander Vasily Tsibliyev was trying to perform an experiment with a new docking procedure, as had been requested by Mission Control. The Spektr module, containing the bulk of Foale's scientific experiments and all of his personal effects, was punctured and became depressurised.

Quick thinking by the crew saved the station, and the Spektr module was isolated from the rest of Mir. In order to close the hatch to Spektr, however, they had to cut the power lines from Spektr's solar arrays to the rest of the station, throwing the crew into darkness and shutting down nearly all of Mir's onboard systems.

Rather than panic, or become frightened, or blame the Russian Commander for the accident, Foale devised a plan to help regain the space station's capabilities. More importantly, he realised that the camaraderie he had established with his Russian crewmates through his previous weeks on Mir would be critical in bringing them through this crisis.

Before the collision, Burrough relates, Foale had set up an impromptu theatre in the Spektr module. Every Saturday night he selected an American film for all of them to watch on a computer monitor. As there were no Russian subtitles, Foale did a 'running translation, complete with voices', according to Burrough.

After the collision, the theatre and most of the movies were destroyed. Both Foale and his crewmate Alexander Lazutkin realised that the Commander, who blamed himself for the accident, needed some psychological relief from the stress of the accident and its aftermath. They decided to try to resurrect the Saturday night movie tradition to raise his spirits. On the Saturday night after the collision, when 'enough power has returned to the station... Commander Tsibliyev decides it is time for all three men to take a much-needed rest,' Burrough reports. Foale selected the film *Apollo 13* – a story of a space mission in which the crew is in much more imminent danger than the Mir crew watching the film. Tsibliyev later recalled: 'That film, it is the best of the best... Everything in the film is so realistic, so truthful, so dramatic, it's just perfect.'

Figure 2.14. The culprit in the collision with Mir on 25 June 1997 was an unmanned Progress supply ship, similar to the one shown here displayed in Moscow in 1981. The Progress is used to deliver supplies to Mir, and to dispose of garbage as it is sent to burn up through the atmosphere. (Courtesy Mark Wade.)

Figure 2.15. This photograph of the damaged solar array on the Spektr module was taken by Space Shuttle *Atlantis* in October 1997 after its docking mission with Mir. The Russian crew and astronaut Dr Michael Foale spent a good part of the rest of their mission trying to stabilise Mir after its collision with the Progress.

Afterwards, all three men agreed that their circumstances were far less dangerous than those of the Apollo 13 astronauts. 'We felt that, especially from a psychological point of view, their situation was much worse than ours,' remarked Tsibliyev. 'We at least had a spaceship [the Soyuz, docked at all times to Mir] that could get us home. With Apollo 13 they had to fly all the way around the Moon in order to get back to Earth.'

Later in the mission, Foale received, via e-mail, a copy of an article that Apollo 13 commander Jim Lovell had written, comparing the two flights. 'Jim said that 'I understand how these guys feel up there, because I've been there as well. I know their courage and bravery.' It was quite wonderful to hear that from Jim', Tsibliyev recalled.

2.10 RESCUING A SPACE STATION

The Russians and the Americans had distinctly different reactions to the collision aboard Mir. Burrough reports that NASA Phase I leader Frank Culbertson, and his deputy Jim Van Laak, both thought that the joint programme was finished. 'They were ready to give up.' The Russian response was summed up by former cosmonaut and Russian Phase I director Valeri Ryumin: 'You have to remember, we've been operating this station for a long time, and believe it or not, we've had a lot of abnormal situations. Our guys are simply tougher than the Americans in these situations.'

Crises aboard Russian space stations, involving loss of power, loss of attitude control, and loss of environmental control, were nothing new with Mir. On 6 June 1985 the Soyuz T-13 spacecraft was launched to Salyut 7, carrying cosmonauts Vladimir Dzhanibekov – who had flown four previous missions and had visited Salyut 7 twice before, and Viktor Savinykh, who had been on the last flight to Salyut 6. As they approached and circled Salyut 7, they could see that the station was slowly tumbling in orbit. Its solar panels were not aligned with the Sun, and the cosmonauts knew that without electricity, the rotating Salyut 7, with which they would be trying to dock, would be a dead, frozen station. They transmitted television pictures to Mission Control, and flight director Valeri Ryumin reportedly described the pictures as 'alarming'.

Four months earlier, Soviet mission controllers had lost all contact with the station. Without ground contact the solar arrays would not stay aligned with the Sun. Without electricity from the arrays, the attitude control system had become dysfunctional, allowing the station to slowly tumble. Without electricity, the thermal control system had stopped functioning, the water pipes froze, and a layer of frost covered the instrument panels as temperatures plunged below freezing point. There were reports in the Soviet press indicating that the station would be abandoned.

Two days after they arrived in orbit, on 8 June, the cosmonauts docked manually with Salyut 7, and found the station in worse shape than they had expected. In his book *Soviet Space Programs, 1980–1985*, Nicholas Johnson explains: 'With no power at all on Salyut 7, the flight plan called for the cosmonauts to retreat to [their]

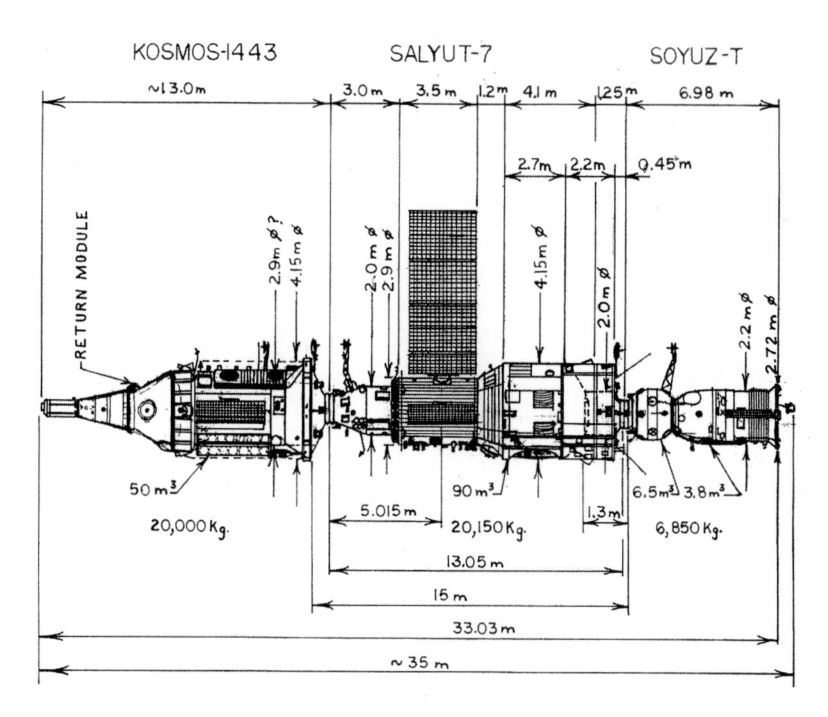

Figure 2.16. Salyut 7, the last in the line of the series of space stations launched by the Soviet Union before Mir. It had only limited access for docked vehicles, but was the first station consisting of more than two spacecraft. (Drawing by Charles P. Vick.)

Soyuz T-13 and return to Earth.' But the cosmonauts were not about to give up that easily. They entered the station wearing breathing apparatus and multiple layers of warm clothing, including woollen hats.

The 'rescue crew' gerry-rigged one of the solar arrays to a storage battery, using cable and objects they found in the station, and began charging it. For several days, Dzhanibekov and Savinykh patiently brought Salyut 7 back to life, charging one battery at a time and returning to the Soyuz T-13 every 40 minutes to warm themselves. The temperature in the Salyut 7 station was estimated at 10° C. Because the thermometers' lower range went down to only 0° C, Mission Control had one of the cosmonauts spit on the wall and time how long it took to freeze, so that they could estimate the temperature!

The crew worked in arctic attire, and reported that their feet became painfully cold. After working without ventilation, the cosmonauts reported that they would get headaches, and feel sleepy and listless from the buildup of carbon dioxide. So they set up a pipe from the Salyut to one of the ventilation systems on their Soyuz spacecraft to remove the CO_2. They were finally able to activate the life support systems in the Salyut station on 12 June.

The cosmonauts restored electricity after replacing cables and bypassing connections that were not working. Equipment that had been damaged in the cold

Figure 2.17. When the first crew arrived, in their Soyuz, at the new Salyut 7 station, they found it in frozen condition. Over a matter of days, with patience and perseverance, they brought the spacecraft back to life. (Courtesy RKK Energia.)

was replaced. On 16 June the station's temperature finally rose above freezing point. As the electrical system was restored, the water onboard thawed, and direct communications between Salyut and Mission Control were re-established. It took ten days to bring the Salyut 7 up to a condition in which mission planners would permit an indefinite stay in the laboratory. Referring to this incident, *Dragonfly* author Bryan Burrough states that the experience with Salyut 7 in June 1985 showed that the Russians could 'almost literally bring a space station back from the dead.'

But despite two years of work onboard Mir, 'the Americans remained largely ignorant of the practical realities of long-duration spaceflight. When there was a mechanical breakdown aboard a Shuttle, the mission was simply ended, and repairs were performed later on the ground. The Russian space stations, of course, never had this luxury. When something went wrong aboard Mir, cosmonauts were forced to fix it in space, which had given the Russians 20 years' experience in the kind of 'seat-of-the-pants' repairs that the Americans had only read about in books. While NASA tended to study and diagram everything to death, the Russians had developed the skills needed to fix things on the run.'

2.11 A CRISIS OF COURAGE

The collision onboard Mir caused a policy crisis in the US space agency, and became a political issue for the opponents of the cooperation with Russia, and particularly for the Republican Party in the US Congress. Michael Foale was scheduled to return from his Mir mission in mid-September, with astronaut Dr David Wolf taking his place. NASA had to make a decision on whether or not it was safe enough to send Wolf to Mir.

On 18 September the House Committee on Science held a hearing on the safety of the Mir space station. In tone, it was more like a lynching. While one would assume that a committee chairman calls a hearing to learn from expert witnesses the answers to questions of import for the nation, it was clear from his opening statement that Chairman James Sensenbrenner, Republican from Wisconsin, had decided what he thought about the safety of Mir before the hearing had even begun. 'What will it take for Russia to decide that Mir has passed its prime or the United States to determine that it's not safe? Does someone have to be killed before NASA and the Russian Space Agency wake up?' Sensenbrenner railed. 'Mir has reached the end of safe operations,' he stated, before anyone from NASA had given their evaluation. The rest of the hearing was orchestrated to try to prove Sensenbrenner's statement.

A more reasoned approach was taken by ranking minority Committee member, Representative George Brown, Democrat from California, who said in his opening

Figure 2.18. The late Congressman George Brown, who was a tireless fighter for the civilian space programme and rational thinking in cooperative space ventures with Russia. (Courtesy Stuart Lewis/EIRNS.)

statement: 'I do not believe that it is appropriate for us as Members of Congress to insert ourselves into the conduct of that [Mir safety] review process. Members of this committee are not in a position to credibly evaluate astronaut debriefings, fragments of engineering analyses, and so forth – as pressworthy as some of the anecdotes that have surfaced may appear to be. We cannot be NASA's safety engineers, and we should not pretend to be otherwise.'

On 16 April, following the February fire onboard Mir, an amendment to House Resolution 1275, the Civilian Space Authorization Act, had passed the Committee. It included a provision which read: 'The NASA shall not place another United States astronaut on board the Mir Space Station, without the Space Shuttle attached to Mir, until the Administrator certifies to Congress that the Mir Space Station meets or exceeds United States safety standards. Such certification shall be based on an independent review of the safety of the Mir Space Station.' The House of Representatives passed the bill on 23 April.

General Tom Stafford, who had been conducting an independent safety review for Administrator Dan Goldin before each Space Shuttle mission to Mir, appointed a special 'Red Team' from among his committee members, to look into the safety of Mir before David Wolf's flight, in view of the series of accidents and equipment failures through the first half of 1997.

At the hearing on 18 September, long-time space analyst Marcia Smith, from the Science Policy Research Division of the Congressional Research Service, presented a balanced and thoughtful view of the situation. On the one hand, she reported that she believed that 'NASA seems already to have achieved most of the objectives set out for the Shuttle–Mir programme,' and she was not, therefore, convinced that additional long-duration flights by NASA astronauts were critical, or would accomplish that much more. However, in referring to the fire and the Progress collision, she said: 'Those emergencies, as undesirable as they were, may have had a positive aspect in terms of demonstrating how the space crews work together in an emergency, how the space and ground crews interact under tense circumstances, and intensified interaction between Russian and American personnel.'

Smith was critical of the way that the media treated the actual situation on Mir: 'It should be borne in mind... that the picture may not be as bleak as what is being portrayed in the media... While Mir is experiencing more anomalies than in the past, as would be expected with an ageing system, the cosmonauts have extensive experience in space station repairs. Mir is Russia's seventh successful space station since 1971... I have studied the Russian space station programme for 22 years,' she continued. 'After seeing them salvage situations that appeared unsalvageable time and time again, it is difficult not to be impressed by the versatility, ingenuity, resourcefulness and determination of cosmonaut crews... So it is not a matter of rejecting concerns about Mir's safety, but more a matter of keeping the newspaper headlines in perspective. As long as a Soyuz spacecraft is available for emergency return, ageing systems alone would not seem to pose immediately life-threatening risks... Despite the many media reports of Mir's imminent demise, the space station continues to function in the hands of its patient, competent crew.'

Smith also addressed the question of the ultimate risk – the death of astronauts in space – by placing it in perspective: 'That there are ten Americans and four Russians who have died as a result of spaceflights indelibly underscores the risks experienced whenever humans venture into space. Despite these risks, the United States and Russia have conducted human spaceflight programmes since 1961... Twenty-three other countries have accepted invitations to send representatives into space on American or Russian missions. Clearly many governments and their citizens are willing to accept certain levels of risk in order to achieve a particular end... the astronauts and cosmonauts who fly into space accept those risks as well.'

During the hearing, Representative Zoe Lofgren (Democrat, California) said that she appreciated the 'bravery of the American astronauts and Russian cosmonauts.' She observed that 'being a congressman doesn't require that kind of bravery.'

During the hearing, NASA Administrator Dan Goldin was threatened by chairman Sensenbrenner that if anything happened to astronaut Wolf on Mir, should NASA decide to send him, Goldin would lose his position at NASA.

At a press conference on 25 September, the day of the STS-86 Space Shuttle launch to Mir, retired Air Force general and former astronaut Thomas Stafford reported that his independent safety review team had just concluded a ten-day stay in Russia, where 'we reviewed the status of each individual system' on Mir, and 'the operational procedures.' The team concluded that 'productive work can still be done on Mir, [and that] the risk of going to Mir was no greater than it has been before.'

Since the press had made great fanfare of the failure of some of the life support systems on Mir, screaming that these endangered the lives of the crew, Stafford explained that on Mir you can operate even some life support systems to failure, because 'you have nearly five levels of redundancy of oxygen. And that's more than I ever had on Gemini or Apollo."

It will not be possible to keep spares of everything onboard the International Space Station, any more than it was onboard Mir. Critical systems will therefore be changed when they reach the end of their expected lifetime, but non-critical systems, and those where there are redundancies, will be 'operated to failure' and then replaced, as they are on Mir.

During a press briefing on 24 September, the day before the launch of the Space Shuttle to Mir, NASA Administrator Goldin said: 'We have heard the calls of some who would say it's time to abandon Mir. We at NASA, especially Michael Foale, who's in space today, are deeply touched by the outpouring of emotion... However, we know that the decision to continue our joint participation aboard Mir should not be based on emotion or politics; it should not be based on fear. Our decision should be based – and *is* based – on scientific and technical assessment of the mission safety and the agency's ability to gain additional experience and knowledge that cannot be gained elsewhere.'

While all of this political hand-waving was going on in Washington, the Russians were trying to understand the political hand-ringing over sending David Wolf to Mir. Russian deputy flight director Viktor Blagov told author Bryan Burrough: 'We can understand their concern about safety. On the other hand, you can't just go around and scare yourself like this all the time... We have an expression in Russia: 'You can't be a like a bird scared of a bush.' Once a bird gets scared, he is afraid of everything. If you get scared, you should not be working in space. Space presents some danger. There is risk. If you are scared, you should go do something safe, like cleaning streets.'

2.12 TWO KINDS OF COURAGE

Ultimately, the decision whether to send David Wolf to Mir lay at the feet of Administrator Goldin. On 24 September, Goldin made the decision to continue the Shuttle–Mir programme. At a press conference in Washington, he called upon the American people to return to the spirit of exploration that had been the foundation of the nation. He said: 'It is only after carefully reviewing the facts, thoroughly assessing the input from independent evaluators, and measuring the weighty responsibility that NASA bears for the lives of our astronauts, that I approve the decision to continue with the next phase of the Shuttle–Mir programme... This is a decision that all of us at NASA do not take lightly... I will not trivialise the risk associated with human spaceflight and exploration. Like all Americans, I know every time an astronaut travels into space there is risk. When we build the International Space Station we will encounter similar problems and there will be danger. But NASA is ready. We are ready because the review assures us. But we're

Figure 2.19. NASA Administrator Dan Goldin has generally had a good working relationship with Congress, but it was severely strained during the political fight over whether or not astronaut Dr David Wolf would carry out his planned long-duration increment onboard Mir. (Courtesy Bill Ingalls/NASA.)

also ready because it's the right thing to do. We overcome the unexpected. We discover the unknown. That's been our history, and that's been America's destiny.'

At a luncheon during the annual meeting of the American Astronautical Society in Houston on 17 November 1998, the man who had been at the centre of the controversy, astronaut David Wolf, observed after his mission: 'It is easy to cooperate when things are going well. The test is when it's difficult... There was a lot of fortitude in the leadership [of NASA] to say it was safe, when there are inherent risks. This is exactly what the American people want us to do, this programme... [T]here is [only] a vocal minority of critics.'

At the same meeting, Jim Van Laak summarised how each astronaut onboard Mir had to deal with something he had never seen before he went. He said that the collision during Michael Foale's increment on Mir was a 'pivotal moment in the programme. Everything was up to the crew, which did a magnificent job. It was clear the future of the programme was on the line... Courage comes in several forms,' he continued. 'One is in the short term, like the [Mir] fire and collision. Another courage is putting one foot in front of another every day. [We will see] if there is the American courage to tough it out.'

Figure 2.20. The last crew of cosmonauts sadly left Mir on 28 August 1999. It has been empty since then, and the plan is to deorbit Mir and allow it to burn up in Earth's atmosphere. Mir leaves behind an important legacy in living and working in space through international cooperation.

In an interview with Andy Thomas, the last NASA astronaut to live onboard Mir, I observed that his increment on the Russian station was much less 'eventful' than the fire during Jerry Linenger's mission, or the collision on Michael Foale's mission. Thomas replied: 'The reason why it was a lot less eventful was because of the experience that was gained in the flights of my predecessors. We learned how to work with the Russians, we learned how to do a science programme on a space station like that. We learned all about the human factors, and we learned how to get the Shuttle up and back. We learned about the [crew] support structures needed on the ground, and the psychological support and technical support. And so, it was not coincidental that my increment [onboard Mir] went very smoothly.'

Andy Thomas has been working as part of a team of astronauts applying the lessons learned from the Shuttle–Mir programme to the International Space Station. He has assisted with the training in Russia of astronauts who will work on the ISS and must be familiar with Russian hardware, including the Soyuz capsule that would bring a crew back in case of an emergency.

He said in the interview: 'I do think it would have been inconceivable to try to do a collaborative space station with the Russians, without having done the Shuttle–Mir programme.'

3

Growing food in space

Newton opened the next meeting with a situation report.

'Now gentlemen, I must ask you to turn your attention to our material situation,' he began. 'Supplies are steadily running out. We are using the reserves as fertilizer for the plants, but we have not got enough fruit and vegetables to use up all the fertilizer. The ship is too small. We shall have to build a greenhouse on to the rocket. One advantage is that it will give us more room to walk about inside without having to put on space-suits. Moreover, we shall not use up our reserves of oxygen and food so quickly – if we have plenty of plants, they will provide us with both. All our waste products and refuse will be absorbed completely. We shall take from the plants as much as we give them. Nor shall we need to keep any reserve stocks; we can dispose of them and satisfy our needs with the carbon and nitrogen substances in the plants. This will be best for us, and even necessary, in view of easy life here, with no heavy work to do and a temperature of 30 degrees.'

Konstantin Tsiolkovsky, *Beyond The Planet Earth*, 1920

This somewhat fanciful description of how man will live in space, and the role of plants in that environment, by Russian space visionary Tsiolkovsky, puts forward the first detailed requirements, as well as engineering details, of how man's nutrition and atmosphere can be supplemented. In addition, scientists, medical personnel, and cosmonauts and astronauts have found that nurturing other living things far away from Earth can be relaxing, and help alleviate the isolation in space.

One array of technologies that will have to be mastered to accomplish long-term space missions will be the ability to grow crops in space. On other solid bodies, such as the Moon and Mars, the task will be somewhat Earth-like, in that a soil is available, and even the partial Earth gravity on these bodies provides growing conditions quite different from those with virtually no gravity at all.

But on space stations orbiting the Earth, Moon or other planets, and in spacecraft making journeys considerably longer than the two days it took Apollo astronauts to travel to the Moon, supplementing the food that is carried from Earth with food that can be grown *in situ* would be a great weight-saving benefit.

In addition, plants can provide a service now performed by mechanical equipment using chemical techniques to 'scrub' the carbon dioxide in the atmosphere expelled when the crew-members breathe. At present, the oxygen they consume has to be replaced with onboard supplies carried from Earth, or the electrolysis of water – also a precious commodity in orbit. Plants – which use carbon dioxide in their metabolism and expel oxygen as a waste product – can potentially take on the task of regenerating the spacecraft's atmosphere.

As important as plants will be for the future of space travel, studying the way plants grow and reproduce without the influence of gravity that effects them on Earth will aid scientists in their understanding of some of the fundamentals of plant physiology. Dr Frank Salisbury, now retired from Utah State University, began working on experimental space farming projects in 1981. The work that Dr Salisbury and his colleagues have been engaged in concerns the understanding of the parameters, such as light, soil chemistry, and response to gravity that effect plant growth, with the goal of maximizing yields in the constrained environment of space.

In 1996, discussing the results of his plant growth experiments on the Mir space station, Dr Salisbury said: 'Even after about 130 years of truly scientific study, we still don't know exactly how plants respond to gravity... Maybe if we see how they grow when there is no directional gravitational force, we can learn something about gravitropsim [how plants respond to gravity].' This will provide insights into one of the many factors that influence plant growth.

In the early Soviet space station programme, attempts were made to grow wheat and peas on Salyut 4 and early Salyut 6 experiments, but they failed at the stage at which flowers should have formed. Andy Salmon reports that there was a debate among scientists about whether there were a 'biological barrier' to growing plants in microgravity – analogous to changes that the human physiology undergoes in the space environment – or whether the technology and hardware were at fault.

In 1980, several orbital 'hothouses' were used on the Salyut 6 station, using artificial soil materials; and in some cases, electrical stimulation was applied to the plants. One facility, Biogravistat, even used centrifugal force as an artificial gravity. In 1980, of the many different varieties of plants tested, only *arabidopsis* reached the flowering stage, but it did not produce seeds. A similar experiment onboard in 1982 produced both flowers and then seeds, taking nearly twice as long to do so as Earth-grown plants. Of the 27 seed pods produced, only ten were fully matured, containing about 200 space seeds. Some of the seeds were later used in Earth-based experiments.

With the goal of long-term spaceflight in mind, in 1984 scientific work was begun under the auspices of the Soviet-led Intercosmos programme, coordinated by the Institute for Biomedical Problems in Moscow, for the 'study of the ways and means for use of higher plants, algae, and animals in biological systems for life support of space crews.' The Soviet programmes and their successors have placed great emphasis on developing closed-cycle life support systems, where the water can be recycled, and plants can substitute for, or at least complement, the chemical regeneration of the atmosphere.

As Dr Tania Ivanova, of the Space Research Institute of the Bulgarian Academy of Sciences, explains in a paper presented to the Congress of the International

Astronautical Federation in 1991, during the 1980s a number of scientists from several eastern block countries were at work on the design and development of various experimental facilities to study the growth of plants in the microgravity of space. The goal has been to develop Controlled Ecological Life Support Systems (CELSS), as closed biospheric systems based on the biological recycling of chemical elements, in order to make space travel as independent as possible of supplies from Earth.

Ultimately, a closed system would use the solid, gaseous and liquid wastes produced by the human space passengers to nourish an array of plants, algae and animals, which, in turn, would provide the food and replenished atmosphere for the crew. To further this research, investigators from the Czech Republic developed an apparatus called Hlorella for the study of algae growth in space. In the Slovak Republic, a new facility, Incubator 2, was used onboard Mir to study the development of Japanese quail eggs, and chicken maturation. And in Bulgaria, Dr Ivanova and her colleagues developed a new type of space greenhouse, Svet ('Light'), to study the growth of higher plants in orbit.

One of the first problems to be solved was the development of a substrate to take the place of soil, that was optimised for the conditions of weightlessness. In 1985 the first preliminary soil moisture experiments, called Substrate, took place onboard the Salyut 7 space station, to test the hydrodynamic characteristics in space of a new zeolite material called balkanin. Zeolites are three-dimensional aluminosilicate crystal structures that can lose and gain water reversibly. Balkanin was used because it is a natural zeolite, which might also be found on the Moon and Mars. It is 'dressed,' or doped, with mineral salts to provide nutrition for the plants. The test onboard Salyut 7 was to verify a mathematical model that had been developed of the water transfer in the substrate medium under microgravity conditions. This experiment was the first time a sensor to measure the water content of a closed substrate volume was tested.

A series of greenhouse modules was being designed and tested on Earth. Dr Ivanova reports that between 1984 and 1987, equipment was tested in a special

Figure 3.1. Dr Tania Ivanova, the head of the Space Biotechnology Department of the Space Research Institute, Bulgarian Academy of Sciences, who developed the Svet greenhouse, is here seen tending onion plants during a ground-based experiment. (Courtesy Dr Tania Ivanova.)

environmental test chamber in Bulgaria, to develop the technology and computer control programs that would be used in Svet.

A major advance was made in June 1990, when the Svet greenhouse was launched to the Mir space station inside the Kristall laboratory module. On 15 June the first experiments with the Svet, termed Greenhouse 1, were begun.

In 1995, under the Shuttle–Mir joint programme, a Gas Exchange Measurement System (GEMS) from the United States was added to the Svet greenhouse. It enhanced the ability to precisely measure the critical environmental parameters affecting plant growth, particularly the up-take of carbon dioxide and the substrate moisture.

During the Shuttle–Mir joint science programme, Greenhouse 2 and Greenhouse 3 experiments were performed, first growing Super Dwarf wheat, and then a plant in the mustard family. Because not all of the experiments have been successful, scientists are learning how growing crops in microgravity differs from agriculture on Earth, and are developing ways to solve the problems that the space environment presents. These insights also allow the discovery of some of the basic mechanisms of plant growth, laid bare in the absence of gravity.

3.1 EARLY EXPERIMENTS IN SVET

The Svet greenhouse is a small facility with a plant-growing area of about one square metre. The root modules, where the seeds are planted, are arranged in four rows side by side, mounted on rails and fitting like a drawer into the bottom of the structure of the greenhouse. There is room for plants to grow to a height of about 40 cm. On the top is the light unit, which has twelve flourescent lamps. The paint on the inside of the plant growth unit was specially developed to be reflective, but contains no noxious chemical elements.

Because of the low moisture capacity of the substrate granules, water is injected into the balkanin from tubes through a foam, which is surrounded by folded pieces of a chloride-impregnated fabric, or wick. The wick increases the surface area for the water to move into the substrate, and also prevents particulate matter from the substrate from escaping into, and floating around, the cabin of the spacecraft. From the beginning of the experiments, plant physiologists working on the project recognised that the major challenge to growing plants in microgravity would be the even distribution of water to the roots of the plants, because water introduced from the top of the soil will not naturally drip 'down'.

The plant growth unit contains sensors to monitor the environmental parameters for plant growth, such as the temperature and humidity of the air and the substrate in different parts of the growing area. The separate control unit receives and processes the data from the sensors, and controls the light, ventilation and water-pump systems. The data are sent three times daily to the ground, during scheduled radio communication.

Growing healthy and robust plants depends upon a delicate balance between the various activities which must be carried out by the plant. The transpiration rate of

Figure 3.2. The Svet ('Light') greenhouse was developed as a step towards both growing food for long-term space travel, and to provide the oxygen from plant metabolism required in the life support systems for human space travellers. There is room inside for plants to grow to about 40 cm. Illumination is provided by twelve flourescent lamps on the top of the greenhouse unit.

the plants, or gas exchange (taking up carbon dioxide and expelling oxygen) bears a linear relationship to the amount of water that the plants can extract from the root zone; too little or too much water disturbs the 'breathing' of the plant. In turn, the transpiration rate determines the rate of accumulation of dry plant matter, or the growth rate. Water is also important for cooling the plants, especially under the intense light of an artificial growing environment. Doubling the light level for the plants, to increase their growth rate, will also double the amount of water that the plants must extract through their roots.

The Greenhouse 1 experiments were performed in Svet after it arrived on Mir in 1990. Dr Ivanova reported in papers presented in 1991 and 1992 that the first fresh root-vegetables – white-ended red radishes and Chinese cabbage plants – were produced in space during these experiments. It was observed during the two-month experiment that the vegetables accumulated biomass in direct proportion to the duration of the light period, and that it was not indispensable to simulate 'night' and 'day'. However, the duration of the light period for the plants during Greenhouse 1 experiments was limited, by the power supply available on Mir, to 16 hours per day.

The scientists reported observing a delay in the stages of development of the space plants, due to difficulties in substrate moistening. In addition, when fresh 23-day and 54-day and dried 29-day roots were returned by the crew for study by scientists, it was observed that the Svet plants were half the size of the ground control plants

Figure 3.3. This close-up view of wheat growing in the Svet greenhouse makes visible the wick material that is placed on top of the zeolite 'soil' in which the seeds are planted. There is room for about thirteen seeds in each of four rows, and the major challenge is to ensure that the seedlings have enough, but not too much, water.

that had been grown under the same lighting conditions on Earth. For example, the Chinese cabbage and radish plants in space on the 23rd growth day corresponded to the ground vegetation on the 10th or 11th day of growth.

From these characteristics Dr Ivanova and her colleagues concluded that photosynthesis had been reduced fourfold in the space greenhouse. She reported at the annual conference of the International Astronautical Federation in October 1997 that physiological and chemical analyses showed that the space plants were 'exposed to significant moisture and nutrient stress'. It was clear that more efficient real-time monitoring of environmental conditions would aid in altering watering procedures more rapidly when they were not optimal, thus reducing stress on the plants.

3.2 GREENHOUSE 2 WHEAT EXPERIMENTS

In May 1995 the new Spektr module was launched to Mir, and on 30 June the Space Shuttle *Atlantis* docked with the station, carrying a new vegetation module for the Svet, developed in Bulgaria. A new light unit was delivered to Mir, to improve brightness, increase reliability of the equipment, and reduce the energy consumption of Svet's illumination system. Also onboard was the Gas Exchange Measurement System (GEMS), designed and built by Utah State University. GEMS was designed to monitor the environment in the Svet greenhouse and in the Kristall module of Mir.

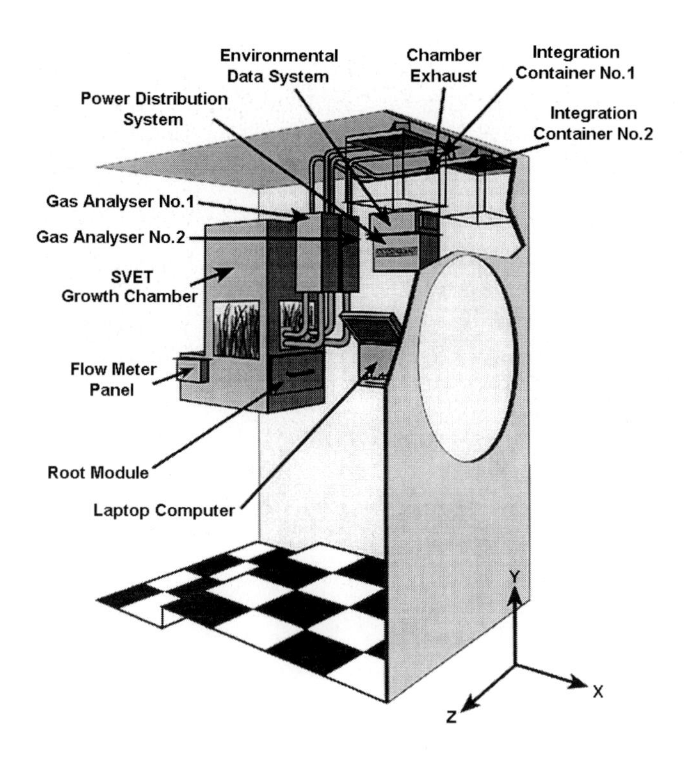

Figure 3.4. The addition of the Gas Exchange Measurement System to the Svet greenhouse during the Shuttle–Mir programme allowed a more precise monitoring of the humidity of the soil, one of the most important parameters that determine plant growth. (Courtesy Utah State University.)

GEMS includes a set of sixteen moisture sensors that monitor the water distribution in the substrate of the greenhouse. The unit measures and records data every three seconds: substrate moisture; light, leaf, and air temperatures; oxygen; and cabin air pressure and temperature.

In a dry run for the new GEMS equipment onboard Mir, before the official Greenhouse 2 experiments began with the participation of American astronauts, from August to November 1995, Dr Frank Salisbury, of Utah State University, led a wheat experiment with Svet. In July 1996 he reported that 'many things went wrong ... two thirds of the lights stopped functioning very early in the experiment', which was quite serious, because even when *all* the lamps in Svet are functioning, 'there is just enough light for food growth.'

While there were other problems, the plants stayed alive for 90 days, Salisbury reported, but 'not a single plant formed even one wheat head!' This was certainly unexpected, as wheat had formed heads in space before, 'so the fact that our plants did not flower is perhaps the most interesting thing we learned in this experiment. We are doing many ground experiments to try to find out why the plants stayed vegetative.'

Figure 3.5. Dr Shannon Lucid, the third astronaut to live onboard Mir, enjoyed her task of monitoring and documenting the progress of the wheat crop planed in the Svet greenhouse, but it was John Blaha, her replacement on Mir, who actually harvested the crop.

The Greenhouse 2 experiment was conducted on Mir between 10 August 1996 and 17 January 1997. On 10 August, astronaut Shannon Lucid planted Super Dwarf wheat seeds in the Svet, with the goal of completing the first 'seed-to-seed' full plant cycle. All of the active components of Svet, including a new fluorescent light, were replaced for the experiment. But the day after the planting, a failure in the control unit required that for the next 90 days the experiment was guided from Earth.

In her article in the May 1998 issue of *Scientific American*, Dr Lucid described the experiment, explaining that 'every day we photographed the wheat stalks and monitored their growth. At selected times we harvested a few plants and preserved them in a fixative solution for later analysis on the ground... One evening, after the plants had been growing for about 40 days, I noticed seed heads on the tips of the stalks. I shouted excitedly to my crewmates, who floated by to take a look.'

On 16 September, Dr Lucid was replaced on Mir by astronaut John Blaha, to whom fell the responsibility of harvesting the dwarf wheat crop. The plants from the Greenhouse 2 experiment developed leaf areas and biomass similar to the control plants on the ground, although they grew a shortened 'crab grass' style canopy. On 12 December 1996, NASA announced that Blaha had harvested the first crop of healthy plants grown in Svet that had completed their life cycle from

Figure 3.6. On 16 September 1996, John Blaha replaced Shannon Lucid on Mir. Here he is seen testing his Russian space suit upon joining the Mir crew. On 12 December, NASA reported that Blaha had harvested the wheat.

seed to seed. The press release from NASA stated that project scientists were optimistic that the plants 'appeared to have matured fully to produce the desired seed-containing heads.'

'Harvest of this wheat on Mir represents the first time that an important agricultural crop and primary candidate for a future plant-based life support system has successfully completed an entire life cycle in the space environment,' co-investigator Dr David Bubenheim, from the NASA Ames Research Center, said. 'The development of plant-based regenerative life support systems is critical to sustaining a crew during long-duration missions such as Mars exploration.'

Dr Frank Salisbury, Principal Investigator for the experiment, added: 'Completion of a plant life cycle in microgravity would prove that there are no 'show stoppers' – no stages in the life cycle that absolutely require gravity for completion. Based on first-hand reports and videos of the plants growing aboard Mir, it appears that our Super Dwarf wheat plants have achieved that critical goal.'

But on 21 February 1997, David Morse reported that scientists found that after the Space Shuttle STS-81 mission brought back the dwarf wheat plants (and Blaha) from Mir in January, the wheat heads lacked seeds. All of the 296 heads that formed on the plants were sterile, and had stopped developing at the pollen development stage, while the rest of the plant parts were normal.

Some in the press immediately branded the experiment a failure, but co-investigator Dr Bubenheim said: 'These conclusions are not justified by the results of the Mir experiments. The potential failure of the wheat plants to set seeds in space was anticipated in advance, since it is known that head sterility can be induced on the ground by several environmental stresses.' More importantly, he stated, 'the samples and environmental data returned from Mir had been designed specifically to provide a diagnostic ability that will be used to aid in identifying why seeds were not produced.' Of course, he added, the fact that seeds were not produced by the plants 'is disappointing, but it does not take away from the major accomplishments of the project.' The task ahead was to find out what had caused the problem.

Ground studies were initiated at Utah State University and NASA Ames Research Center to replicate the conditions on Mir to try to isolate the cause of the sterility of the plants. In a presentation in November 1998, Dr William Campbell, of Utah State University, reported that scientists exposed wheat to concentrations of ethylene gas on the ground similar to those on Mir (about 300 times the normal allowance), and found that they repeatedly exhibited responses similar to those observed in the Mir plants. They concluded that the exposure of the wheat plants to 1–2 parts per million of ethylene gas had led to small and shrivelled pollen grains and sterility.

Dr Campbell reported that the scientific literature 'is replete with the use of ethylene to induce male sterility in cereals', which provides a strong case for their conclusions regarding the data from Mir. Dr Mary Musgrave explained in an interview that ethylene is a growth regulator hormone for plants, and many are more sensitive to it than are people. For better results, she suggested, a scrubber system would be needed to remove excess ethylene gas.

Dr Gail Bingham reported in a paper delivered in 1997 that the moisture control in this experiment was far better with the new automated GEMS system than had been achieved on Mir the previous year. Scientists were able to quickly adjust the amount of water reaching the plants before stress could set in, and the water level could be maintained close to 'ideal'. Scientists who have laboured for years to understand the physics of water and trapped oxygen movement (bubbles) that can effect plants in space, gained new insight from the Greenhouse 2 experiments.

But there loomed one large question: What would happen to the mustard plants that were to be planted on Mir by astronaut Michael Foale in the follow-on Greenhouse 3 experiment?

3.3 THE FIRST SEED-TO-SEED IN SPACE

When Foale launched on the Space Shuttle to Mir on 15 May 1997, he became the fifth astronaut to live on the space station. The British-born scientist joined NASA in June 1983, and four years later was selected as an astronaut candidate. Before his stay on Mir, Foale had been a crew-member on three Space Shuttle missions. Currently (late 1999) he serves as Assistant Director (Technical) at the NASA Johnson Space Center, and is assigned as a crew-member on the third mission to service the Hubble Space Telescope.

Figure 3.7. As can be seen in the photograph, the wheat grown in Svet appeared to be normal and mature, with numerous heads on the tops of the plants. The scientists (and astronauts) assumed that they contained the seeds for the next generation of plants. They were sorely disappointed to find, upon their return to Earth, that the plants were sterile.

Riding with Foale aboard *Atlantis* on the way to Mir were *brassica rapa* seeds, to be planted in the Svet greenhouse. Dr Mary Musgrave, of the Louisiana State University Agricultural Center, the principal investigator for Foale's experiment, along with Dr Margarita Leviniskikh, of the Institute for Biomedical Problems in Moscow, describes *brassica rapa* as derived from the economically important mustard family. This plant was chosen for this space experiment because of its short lifespan, low light requirement and small size.

Prior to Foale's launch to Mir, Dr Musgrave explained that the goal of the experiment was to study 'a plant's entire life cycle in microgravity and better understand the effects of microgravity on plant reproduction'. The plan was for three successive seed plantings in Svet, 'starting with dry seeds grown on Earth, planting them in space to get a second generation, and using the resulting seeds to get a third generation of plants.'

On Earth, the plants take about 14 days to flower after planting, and the plan was for Foale to then spend seven days using a bee stick to transfer pollen from one plant to another. After 45 days the plants were to be harvested, and the seeds from the first generation would be planted. Prior to launch, Dr Foale remarked, 'I never was a farmer in my childhood, but I'm ready to be one now. I'm a physicist, in fact, by training, but this idea of growing seeds from a seed to a fully flowering plant... pollinating first, and then collecting the seeds, kind of excites me. I'm looking forward to that.' At the end of his second week onboard Mir, he had completed assembly of the Svet greenhouse, and was ready to plant his crops.

In an interview in his office at NASA's Johnson Space Center, Houston, on 13 May 1998 Foale described his Svet experiment: 'The goal of the investigators on my experiment... was to grow *brassica rapa*, and to take that from the seed, all the way through flowering, through pollination to seeds in a pod, and then harvest the seeds, and then replant them... The hope was that in my time on Mir, which would be four and a half months, I would do this three times, and produce two space generations of seed, from which, in turn, would be produced plants in space. The goal was, specifically, to attack the problem of not just germination, but pollination and then the production of seeds. And then fruition and harvesting, and then repeated plantings of those space-produced seed... Basically, we were successful.'

The main problem in Svet was to try to improve the delivery, distribution and absorption of water to the plant roots, to control the moisture content of the simulated soil. Over the years the Bulgarian and Russian scientists had experimented with various substrates. Foale reported that more than the substrate material was important: 'The root module was a box that contained zeolite powder, but it is not a uniform powder. It has small grain sizes and large soil sizes. And this is very critical, they have found in space research, in getting the water to not flood the root or totally dry the root. You want the water to form patches, but not to totally flood the zeolite when you feed the water in. There's no gravity to differentially separate the water from the root.' In fact, the zeolite substrate has a range of granule size between 1.7 mm and 4.3 mm.

As we who have grown plants are aware, 'on Earth, in soils, when the water goes in, it's more dense as you go down, and it's drier as you go up. So, the root can find places where there's air, and places where there's water, and pick or choose the environment... In space, for many, many years with the Svet module, they had very little success, because the soil material didn't work out. It was always a fine powder of one particular size, 70 microns. And that would either totally wet and flood the root, or totally dry out the root. But either way, they couldn't get viable plants. I think a big breakthrough was this zeolite.'

Another important breakthrough was from 'the University of Utah, for coming up with a system that allowed this wetting to be done in a more root-friendly way,' Foale reported. They 'came up with a much more sophisticated system of measuring the humidity, both at the surface and at depth in the soil. But even though that was sophisticated, and corroborated the rather simpler measurements that the Bulgarian-built Svet measured in the soil, nonetheless those simpler measurements were the ones that were used to control the humidity level. They confirmed that just having one or two sensors, as they originally planned with the Svet module, is good enough. It was what level to control to that they didn't know.'

'In the early days, people always tended to over-wet the root module,' Foale explained. That can still be a problem. 'I think now, on the third planting, I remember I got distracted, and I ended up leaving it on high wetting, for about four or five hours longer than I should have. And when I came back, I thought, 'Oh, no – overshoot!' If I wanted to I could have told the ground that everything was okay.

But I knew that pretty soon we'd have an overshoot, and then even though I would have turned off the water, the humidity level would slowly increase over a day or two. So I discussed it with the ground, and we just agreed to keep the lights on a bit longer, because that also warms, and it tends to dry out the soil. The lights are what determines the major water loss of the soil.'

3.4 FARMER FOALE

Dr Foale is able to provide a unique 'hands-on' description of how the Greenhouse 2 experiment was carried out during his increment on Mir. For the plantings, he explained in the interview, 'we had root modules prepared. A root module allowed four rows of plants watered in pairs, to grow about 13 seeds per row. So, we could do about 52 plants at a shot. That was the plan: 52 seeds planted.'

Foale described his efforts to control the most critical parameter of the experiment – the amount of water being delivered to the roots of the plants: 'On top of the soil material, wicks are placed, that spread out down into the zeolite material. I would place this root module, which is roughly a half-metre by a half-metre, into the greenhouse, and connect up water tubes and electrical sensors. I would measure the temperature, along the strips, at different depths in the soil, as well as free water. The whole point of this was to measure the humidity, the water-level content, at different points in the root module, near the plants, so that the machine itself could control the wetting, and it didn't flood the roots or dry them out... The first day I set up the equipment', he reported, 'which took a lot of time, finding all the pieces that were left over from the expedition before mine... I set up all of the electronic monitoring of the temperatures of the soil, and then took out long cellophane strips on sticky tape of the seeds. The round seeds were about a millimetre in size, and I would place them, by hand, with tweezers, into this wick, about 13 per row.'

Then, the most tricky part of the job 'was to wet the zeolite and the root module enough, but not too much. And initially, on the first try, I think we overshot. We weren't getting any measurements of humidity in the first twelve hours or so, so I and the investigators agreed that we'd put more water in. And lo and behold, suddenly, we saw the humidity levels going way past 50%, and we were afraid we were going to flood the root module.' But that was only the first day, Foale reported. 'We got better at it the second and the third time around. At the same time, we left the light bank on continuously, in a cycle of 23 hours on, and one hour off.'

Once the astronaut and the scientists on the ground had become more adept at watering the seedlings in this unique environment, the seeds started to germinate. But, as Foale stresses, 'in weightless conditions, there is no up or down.' About 50% of the seeds 'put a little root shoot – the up-growing shoot on Earth – burrowed down into the wick, and the root would start popping up. It would be completely confused. About another 50% would go heads up and tail down, which is the right way we wanted them. And others would just grow [horizontally] along the wick. This was especially true on the first planting, because I had placed the seeds fairly deep

Figure 3.8. The Greenhouse 3 experiment in Svet was conducted by astronaut Dr Michael Foale. Rather than wheat, however, the seeds he planted were of *brassica rapa*, which is derived from the mustard family. Before his flight, Dr Foale posed for this photograph at the Institute for Biomedical Problems in Moscow, with the ground duplicate of the space greenhouse, in which control *brassica rapa* plants are growing. With Dr Foale are Keith Zimmerman and Sally Greenwalt from the United States, and on the right is Dr Margarita Levinskikh from the Institute, who, with Dr Mary Musgrave in the US, was a Principle Investigator on this experiment. (Courtesy Dr Mary Musgrave.)

down into the wicks, so there wasn't much light getting down to them. The wicks were made of a whitish material, so some light got down.'

Indicating the care he gave the greenhouse, Foale continued: 'When they finally popped their heads up above the wick – or I teased them up with tweezers so they could see the light – then the light, through phototropic action, drew the plants upwards in the right direction, toward the light, and the roots, in the opposite direction, went down toward the wick... I think almost all of the seeds germinated. I'm not an expert; 'germination' to me meant they'd burst their little shoot out. Beyond that, some of them failed. And only about 80% would actually carry on – again, with me helping them in the right direction – to go up toward the light, and reach down... In a matter of a week or two weeks they would start to grow one or two primary leaves, and then flower buds. And only after about four weeks did the flower buds – now the plants would be about five centimetres, six centimetres high – suddenly produce a plethora of yellow flowers, which had pollen on the stamens... I would write down every day the stage of the plants: how they looked, how many had buds, how many didn't. And I would write down the average highs, and then the minimum high, and the maximum high for each row. And, of course, at the same time we'd carefully monitor the temperature and the humidity level in the root module.'

As the seedlings were growing, ambient air from the station was being blown across the greenhouse by the fans that circulate the air for the crew. In order to make precise observations of the atmospheric environment of the plants, and observe how that would change over time, indicating the health of the plants, measurements had to be taken of the gases in the atmosphere around the greenhouse.

'In the first experiment, before the collision of the Progress cargo ship and Spektr,' Foale reported, 'after the plants had grown about three or four centimetres we placed what we called leaf bags – just basically polyethylene bags that were sealed – over the plant. The gas that would go in from the cabin, into the greenhouse, would be passed over the plants, and then passed out through a gas analyser. A measurement was made of the carbon dioxide intake and outtake, the respiration rate of the plants during most of the growing phase.'

Then would come the stage in the growth of the plants that took the most skill. 'It was only when I was going to pollinate that I would take the leaf bags off. And I would then take a bee stick – pieces of bees on sticks. This is actually a common technique. I didn't know this. In schools – most schools know about this, teaching high school kids how to pollinate plants – they find bees, and they chop the tails off, and stick them on sticks, on little toothpicks. And you basically go along with your toothpick, with the bee on the end of it – your bee stick – bzzz, bzzz, bzzz – you go up and down the rows... I practiced some of this on the Earth, and I did it pretty well in space, it turns out. You collect pollen first from the stamens, and then you put the pollen onto the pistils. And after collecting pollen in one pass up and down the row, like a bee would, you then go into two passes, trying to take the pollen you've collected, and put it onto the female part of the plant... There's only a day's window, or two days' window, in this plant's very rapid growth cycle, when any one flower is ripe to be pollinated. And it's not quite the same time as when it's ripe to give pollen. So, we do the pollination over a period of about a week, about four weeks into any one planting.'

When pollinating is finished, 'you put the leaf bags back on so you can measure the CO_2 over the plants again. And another two weeks went by, which was already two weeks longer than we expected. We thought the cycle would be four weeks. It is on Earth. It takes longer in space.' By this time the collision with the Progress ship had occurred, and, Foale explained, 'conditions were harder on Mir. The temperature wasn't quite so constant; it got cold a lot in that [Spektr] module. So, it slowed down things.'

After six weeks, 'some pretty long seed pods – just like pea pods – grew in the place where the flowers were. And it was pretty clear that they were full of seeds. They seemed to be full of seeds. And when I was told finally to harvest, I just took down the whole experiment and collected the plants in a glove box. This was invaluable stuff.'

Harvesting the small plants turned into a test of patience, and also innovation. 'In any spacecraft, your whole life is dependent on air flow moving around, blowing the carbon dioxide away,' stated Foale. 'And yet, when you pull out these flimsy little plants that are so light and drying now, they'll blow away in a heartbeat. And so I had to be very careful in collecting these, especially the seed pods, and popping them

Figure 3.9. When the wheat plants that were returned from Mir were found to be sterile, scientists suspected it was a concentration of ethylene gas which caused the fault. Ground tests – the results of which are seen here – confirmed that hypothesis. Both Dr Foale and Dr Musgrave were hopeful that the gas would not effect the *brassica rapa* plants, and they did not.

into little vials with a desiccant. At the same time I took the rest of the plants and put them into a general container bag, so I could then put them into formaldehyde to fix them, so the product could be studied on Earth.'

This was a job that was intricate and laborious. It took farmer Foale three days or so to harvest the crop. The seed pods containing the seeds to be replanted were dried for about 50 days. Foale then stretched a sticky tape across the base block of Mir, and started, very carefully, taking these crispy dried seed pods between the piece of tape, 'and breaking the tape, wiggling it, so that the dry, husky material would fall apart, and just the seeds would be exposed'. These were tiny seeds, half a millimetre. These were *smaller seeds than Earth seeds*. They were not as strong. Some of them would float away very, very quickly. And I'd try and catch them, and then stick them down on the tape that I had for collecting them.' Half of the space-grown seeds were replanted. Each row would have half space seeds, and the other half original Earth seeds, which 'I had a supply of, also on sticky tape. And then the whole experiment was repeated.'

Circumstances were about to quickly change onboard Mir and for Svet. 'I think it was about the time I'd repeated the experiment, that we had the collision,' Foale recalled. 'And unfortunately, the plastic bags that would go over each root module to measure the carbon dioxide intake and outtake – I had stored those in Spektr. So they weren't usable. But actually, for me, it ended up being an easier process, because then I could have access to the plants every day. I didn't have to peer through some kind of crinkly material to figure out what was going on. And if a plant was kind of going crooked, I could redirect it more easily.'

Not only were some of Foale's supplies gone after the collision, the Spektr module spent hours in the dark and the cold. Even before the accident, parameters such as the level of carbon dioxide – food for the plants – could not be controlled onboard Mir. 'In this experiment, especially on Mir, the carbon dioxide level does vary quite dramatically,' Foale explained. On Mir 'it's the same atmosphere [in the greenhouse] as what we're breathing, and the carbon dioxide [levels] went through a large range, so you wouldn't be able to pick out, I think, the effect of carbon dioxide on the plants... Also, the temperature variations in that [Spektr] module were very great. After the collision, in that time frame and afterwards, the temperatures were down even in the 5° Celsius range, really cold in that module, when we really wanted to be running at about 20–25° Celsius.'

Work on Mir, in general, certainly was more difficult after the collision. Foale reports: 'That second planting was a very slow planting, because that was when

Figure 3.10. 'Farmer Foale' found comfort and satisfaction in tending to the seedlings that grew to maturity and produced the first successful seed-to-seed experiment in space, particularly after the Progress collision with Mir and the loss of the Spektr module which contained his personal belongings and science experiments. Here he is exercising on the treadmill in Mir just before returning to Earth, on 30 September 1997.

we were having our most trouble recovering from the collision, and that module was totally unpowered. The only thing in the whole of the Kristall module that had anything on was the Svet, because I was running the power to the Svet, all the way with a long extension cord, from the base block. So it was a pretty hard environment for that experiment.'

But after the accident, farmer Foale had more time and attention to devote to the greenhouse, since other of his experiments were in the unusable, damaged Spektr. He also found ways to improve his farming. 'In the second planting I developed a number of techniques that were more efficient than the first time. Certainly, I figured out how to set the seed, with tweezers, into the wick, so that they didn't float away. The first time I put the tweezers in, opening up the wick with the tweezers, to drop the seed in, but then the seed would just fly out. So I had to figure out a way to put the seed in so that it would stay. I basically ended up being able to put the seed in a little more shallow depth. And as a result, the seeds saw the light quicker, and more of them found their way up than on the previous planting.'

But he felt that the results from the first planting of space-grown seeds was disappointing. 'Only three or four of the space-produced seeds germinated, out of the six or so I planted. We produced about 15 or 20 seeds total – space seeds. But they were so weak and flimsy, only two or three of them were worth even planting. I planted a total of six or seven. And I think only two of those actually ended up producing a viable plant that grew up.'

In general, the second-generation plants were not as vigorous as the first, and were smaller. 'I think the uncertainty of how the seed gets the nutrient through the wick is enough to always favour the seed that's bigger, to do better. It really matters how big the seed is. How much built-in carbohydrate material that seed has, determines how far it gets going before it has to stop pulling from the wick. If you had a very nutritious soil and environment, then it wouldn't really matter too much how strong the seed was.'

3.5 A MILESTONE IN SPACE PLANT BIOLOGY

In the midst of Foale's farming, on 25 June an unmanned Progress supply ship collided with the Mir space station, decompressing and rendering useless the Spektr module. The consequences of that accident led to problems that stressed both the crew and Dr Musgrave's mustard plants, including changes in the atmospheric pressure and composition and three days of darkness for the tiny plants, just as the seed pods from the initial planting were maturing.

Dr Musgrave reports that both Foale and the plants rose to the occasion. As Musgrave reported on 1 August 1997, for the first time, seeds that were produced in space had been planted and germinated. 'This is really a historic time for us,' she said. 'This seems like a real milestone in plant space biology. My first reaction was feeling really good for the whole group of people who have been involved in this project, and especially good for astronaut Mike Foale.'

NASA reported that 'researchers from the US, England, and Russia, including co-investigator Dr Margarita Levinskikh, of the Institute for Biomedical Problems in Moscow, were delighted about the news.' Dr Musgrave reported that the space plants 'took the same amount of time to flower and produce seeds on orbit as they do under normal gravity conditions. They were able to flower and make new seeds, and this means that all of the basic processes that you would hope for a plant to be able to do can happen in microgravity.'

In a September 1998 report on her experiment, Dr Musgrave explains that it had been observed on previous short-duration Space Shuttle flights that 'identical gaseous environments have very different consequences for plants growing in them on the Earth and in space.' In space there is no convective air movement, and, in the absence of active ventilation, stagnant air layers can form around plants.

But during Foale's stay on Mir he was able to complete two entire growth cycles, Musgrave reports. At the time of the collision in June, the plants experienced 72 hours of darkness, and to assess the effect on the plants, and 'advise astronaut Foale on future operations,' Musgrave and colleagues created the same conditions for their control plants growing on Earth. The ground-based laboratory plants *did* exhibit diminished seed weight, and had a higher percentage of undeveloped seeds than did plants that continued in the light.

There were temperature fluctuations on Mir as the cosmonauts worked to repair some of the damage from the collision, particularly during the hours when the Kristall module housing Svet was closed off, as the crew installed a hatch on the isolated Spektr module to allow them to connect power cables and increase electric power to the station. And there were also carbon dioxide concentrations ranging 15–20 times above normal levels on Earth.

At the end of August 1998 Dr Musgrave reported that her team had just completed a 4½-month ground-control experiment which replicated the day-to-day conditions on Mir that her plants experienced in Svet. Similar to the Svet plants, the ground control plants were smaller than normal.

According to Dr Musgrave, in her discussion with a crew-member of a previous isolation experiment in Moscow, she learned that the psychological impact of caring for and observing plants while simulating a space mission was as important – or perhaps *more* important – than as the contribution the plants made to regenerating the atmosphere or potentially providing nutrition.

A similar experience is reported by Dr Gail Bingham, from the Space Dynamics Laboratory at Utah State University, when debriefing a crew that had spent time in the Russian isolation chamber that simulates space conditions. They reported that they woke up early just to check the plants. They felt they had 'become part of the crew'. The greenhouse experiments, Dr Bingham concurred, 'are extremely popular with the astronauts and cosmonauts.'

This was certainly the case for astronaut Michael Foale. 'I don't want to overstate it, because I believe my time on Mir would have gone pretty much the same, but especially when I had almost nothing else to do for a month or so after the collision, the greenhouse experiment really provided me a lot of peace of mind,' he reported in his interview. 'This whole business of being a gardener, and getting used to your plants, and getting into their condition, and developing some kind of connection with them, though romantic, actually has a little application in space, because it provides a very different visual scene and activity from the normal, extremely technical kind of bare and artificial existence that you have in space. It's a connection with the Earth, I guess, that you've brought with you, that gives you some comfort... So, I very much enjoyed doing my morning status check in the greenhouse. It would only take me about 20 minutes each day, but I'd sit there, and I'd just savour that time. And, I think, on a long-duration flight, or on any space station, not only to provide scientific research, but also to provide psychological support, experiments that grow things can be well utilised – and I mean visible things, like plants. I expect animals would be just as rewarding, but they're more complicated.'

I asked Dr Foale if he would consider taking plants on long-duration missions just to take care of them, and not as subjects for experiments. 'Yes, very much so', he replied. 'I think, just like we have house plants for no reason but for their being there, I think exactly the same – in fact, more so – would we value having Earth plants in space, for no reason but that they're pretty, or that they're a reminder of Earth. It's something to follow. They grow, they flower.'

Considering the long term, he said, 'I think the pragmatic value, of course, is self-sufficiency, away from the Earth, in your food production. And the only way you can achieve that is through some of these pretty aggressive technological programmes to grow biological material in space, or on the surface of Mars. So, these are essential to feature in the exploration of space. Because in the end, it becomes too difficult to supply all that food from the Earth. It would be a terribly burdensome logistics problem just to keep supplying food, because you can't make it *in situ*.'

Before new technology to grow plants is sent into space, it is tested on the ground. 'We have a chamber here [at the Johnson Space Center], a 60-day chamber, where they put three or four people inside for, I think it was 60 days, and then 90 days. And they lived there. It was a totally closed system. I think 25% of their CO_2 was scrubbed by the plants that they grew. Pretty significant. And I think they ate most of their plants. So, they did pretty well. And I think 25% of the O_2 was produced by the plants, which is pretty dramatic.'

Russian researchers and cosmonauts continued the experiments with the Svet greenhouse after the Phase 1 programme with American astronauts had ended. In the Greenhouse 4 experiment, the 'Apogee' wheat variety, developed by Utah State University especially for controlled-environments, was used. This variety is designed for high yield and a short stature. On 20 November 1998, 100 wheat seeds were sown in the greenhouse by the Mir 26 crew. Of those, Dr Bingham reports, 12 plants were produced. Those dozen plants produced 29 heads, which contained 518 seeds. This was the first seed-to-seed experiment with wheat in space.

All of the seeds were harvested on 26 February 1999, with all but ten of them to be returned to the ground. Ten of the space seeds were wet and put into a plastic bag and refrigerated at 4° C for four days. They were then taken out to reach room temperature, and all of the space seeds germinated. They were then transplanted to the Svet greenhouse for experiment number 5, but only one space plant

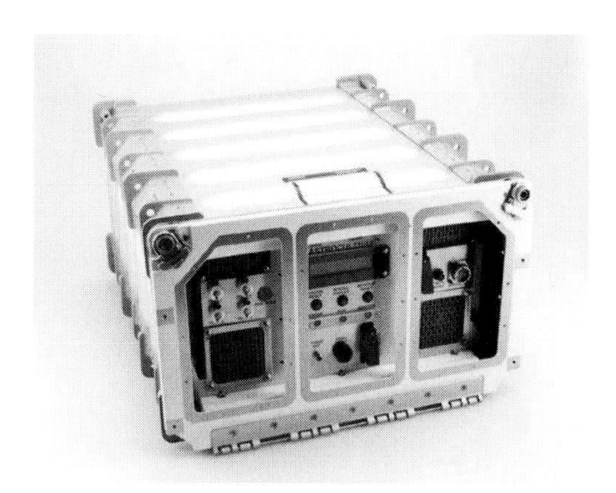

Figure 3.11. The Astroculture facility is a commercial plant growth unit that is being used by industrial and scientific organisations to study the development of plants in microgravity, to understand how to change and improve their characteristics on Earth. (Courtesy Wisconsin Center for Space Automation and Robotics.)

subsequently produced seeds. It is possible that there had been more seeds produced, Dr Bingham reported, but they stood dry from 7 June, when they were harvested, until just before the last Mir crew came back to Earth on 28 August, and may have fallen off.

Dr Bingham and the Russian investigators believe that the surprisingly poor showing for the ten space seeds was due to the fact that the wicks in the Svet had not been cleaned between plantings, and that there was a build-up of fungus and salt. Nevertheless, this was the first successful growth of two generations of plants from space seeds, accomplished during the Mir programme.

Of the approximately 500 seeds produced in space that came back to laboratories on Earth, 89% germinated when planted, and were 'just a little bit shorter with a little lower yield' than the control plants grown on Earth, according to Dr Bingham.

In addition to the Svet greenhouse series of experiments onboard Mir, there was an attempt to use hardware developed for shorter-term experiments on the Space Shuttle to grow food in space.

The Astroculture flight unit was developed at the Wisconsin Center for Space Automation and Robotics (WCSAR), a NASA-sponsored Commercial Space Center, located in the College of Engineering at the University of Wisconsin, Madison. One of its primary areas of work is to develop robotic technologies for plant growth and the genetic engineering of crop varieties with NASA, universities and private companies.

The Astroculture unit flew on a number of Space Shuttle flights between 1992 and 1995, and in January 1998 was flown to Mir on the STS-89 mission, for a seed-to-seed experiment with wheat. It suffered a malfunction in its computer system soon after delivery to Mir, due to a radiation hit to a computer chip.

Later that year, on the STS-95 flight in October, the Astroculture unit was used for two commercial, and, therefore, proprietary experiments, growing a miniature

Figure 3.12. During the Greenhouse 4 experiment in Svet, after the completion of the Shuttle–Mir programme, twelve plants were produced from 100 wheat seeds. Dr Gail Bingham, at the Utah State University, reports that these dozen plants produced about 500 seeds. On Earth, the plants grown from space seeds (right) are comparable to the control plants grown on the ground. (Courtesy Dr Gail Bingham.)

rose flower for fragrance, and a gene transfer experiment with soy bean seeds. On 23 October, before the flight, retiring director Raymond Bula said he hoped that low-gravity conditions would make it easier to introduce new genes into plants. In addition to improving crops, plants are being genetically engineered to produce human proteins, and even vaccines that can be eaten. Now the process is hit-and-miss, he said. 'If we can improve the gene transfer process to even one in 100 being successful, I think there would be tremendous industry interest.' On the STS-95 flight there were about 1,000 soy bean seedlings wrapped in simple water-soaked rolls of paper. The Shuttle crew was to insert a new gene into the seedling to strengthen the immune system. According to Bula, the intention was to induce the soy beans into producing a drug that would relieve arthritis symptoms.

The US plans to have two of high-technology plant-growing chambers on the ISS. The Russians, who are not in a position to develop such expensive hardware, plan to fly a chamber known as 'Lada,' according to Dr Bingham, which is the 'working man's' version of the greenhouse, similar to the car after which it is named. Lada will be smaller than Svet, but with two growth chambers. It will incorporate a module design which can accommodate various experiments, with a separate control unit and the same infrastructure as Svet. Dr Bingham reports that it is being readied for a launch at the end of 2000, and will be placed in the Russian Service Module, Zvezda.

Figure 3.13. In the final Greenhouse 5 experiment, of the 500 wheat seeds produced in space, ten were then planted in Svet, with one plant producing seeds. On 25 November 1999, Dr Bingham reported that all five of the second-generation space seeds had germinated and produced healthy, green plants. In this photograph taken on Mir, the wheat head from a ground seed (left) is compared with a head grown from a seed produced in space during the Greenhouse 4 experiment. Research with the Svet greenhouse has moved us closer to being able to grow food in space. (Courtesy RKK Energia.)

All of the scientists from different nations who have participated in the greenhouse experiments await the facilities that will be available on the International Space Station. As Dr Musgrave states, 'Long-duration access to orbital platforms and the dedicated time of well-trained astronauts will be necessary to develop the database needed to implement the technological goal of a plant-based life support system.'

4

A biorevolution through tissue engineering

For the past hundred years, medical and biological researchers have tried to grow human tissue in the laboratory in order to study the basic biological processes at work in cells – the building blocks of tissues and organs. Scientists have been investigating the biochemical, molecular and, more recently, genetic processes that are the foundation of life.

In addition to improving the fundamental understanding of the functioning of healthy cells, scientists also grow tissue in the laboratory to study the *abnormal* growth of cells, to better understand the process and progression of diseases such as cancer. Researchers have tried to test pharmaceuticals and other preparations for the treatment of disease on human tissue outside the human body, where they cannot inadvertently cause harm due to side effects. There has also always been the hope that artificially grown healthy human tissue could be used to replace diseased or dysfunctional organs which, until now, has been possible only in the imagination of science fiction writers.

Until the recent development of a new, elegant space technology, researchers have been limited to growing tissue in the laboratory in two dimensions, essentially on a flat petri dish or in a flask. In conventional tissue culturing, cells sediment to the bottom of the container as they grow, under the force of gravity. Such a two-dimensional sheet of tissue, in many important ways, bears little resemblance to the same tissue *in vitro*, growing in three dimensions in the human body. Such a two-dimensional aggregation of cells can provide only very limited models of functional human tissue.

The recent invention of the bioreactor – initially used to simulate the condition of weightlessness on the ground – has become a revolutionary new tool in both space- and ground-based research, which is opening a new era in understanding the fundamental mechanisms that govern cell behaviour and tissue formation, and which is already having a significant impact on improving human health.

4.1 SCIENCE IMITATING LIFE

For many years, scientists have been frustrated with the limitations of studying life in a petri dish. Cells grown outside the body do not differentiate or become specialised to perform various functions as they do in the human body. Such two-dimensional *in vitro* cells do not produce the proteins and other biochemical materials that determine the functioning of the cell, and the development of a community of cells into tissue and organs. Scientists have used animal models extensively, particularly to study disease and treatment, but are well aware that there are critical differences between mice and men.

As was stated by Dr Neal Pellis, the Director of the biotechnology cell science programme at NASA's Johnson Space Center, in an interview in August 1999: 'I worked for 25 years modelling cancer therapy modalities in mice. I can tell you that if you're a brown mouse with a tumour on your back, I was the doc in town to see, because I could cure them all. But that's a model. We cannot mechanistically address that same neoplasm in human beings, exactly the same way we could in mice.'

A team of researchers at NASA JSC recognised that if cells could be grown without the influence of gravity, they would not all fall to the bottom of a dish and form a sheet, but would more accurately mimic the three-dimensional growth in a living system. The solution to the problem, they decided, was to have the cells suspended in a growth media, where they could freely aggregate to form tissue more closely resembling that in the body.

Dr Pellis explained to NASA's *Microgravity News*, in the autumn of 1998, that the researchers decided that the first step in the process of developing such a device to take advantage of microgravity was to create a culturing facility that would simulate some aspects of microgravity on the ground, before attempting to design a machine to fly in space. He noted that simply stirring a mixture of cells in their nutrient-filled container could avoid their settling to the bottom, but such an activity is deleterious to the cells. 'Most cells do not like to get beaten around by mechanical shear or hydrodynamic shear', he stated. Mechanical shear would be produced when the cells bumped into the side of the containing vessel as they bounced around, or into one of the propellers used to stir the media. Hydrodynamic shear would occur when the liquid media rolled over the cells during stirring, similar to the action of water rushing over and around rocks. A more 'quiescent environment' was required.

In 1986 three members of the small team at JSC, who were examining these questions, developed the idea of growing cells in a cylindrical vessel – a bioreactor, that would rotate to simulate microgravity. The three pioneers were medical doctor and engineer David Wolf at NASA, Krug Life Sciences bioengineer Ray Schwartz, and Krug engineering technician Tinh Trinh.

In an interview in 1998, after his increment as a member of the crew on Mir, Dr Wolf described the early development of the bioreactor, seven years before he became an astronaut. 'It actually happened when the Shuttle was grounded after the *Challenger* explosion,' he stated. 'We were restricted to doing our research on the ground, and we were attempting to simulate zero gravity.' Years before the bioreactor was conceived, NASA was conducting experiments to grow cells in space. The space programme needed to produce a device to simulate microgravity, in order to protect delicate cells that were being taken to and from orbit by the Space Shuttle. Although cells would be grown in microgravity, the multi-g environment of launching and landing the Shuttle could lead to their breaking apart.

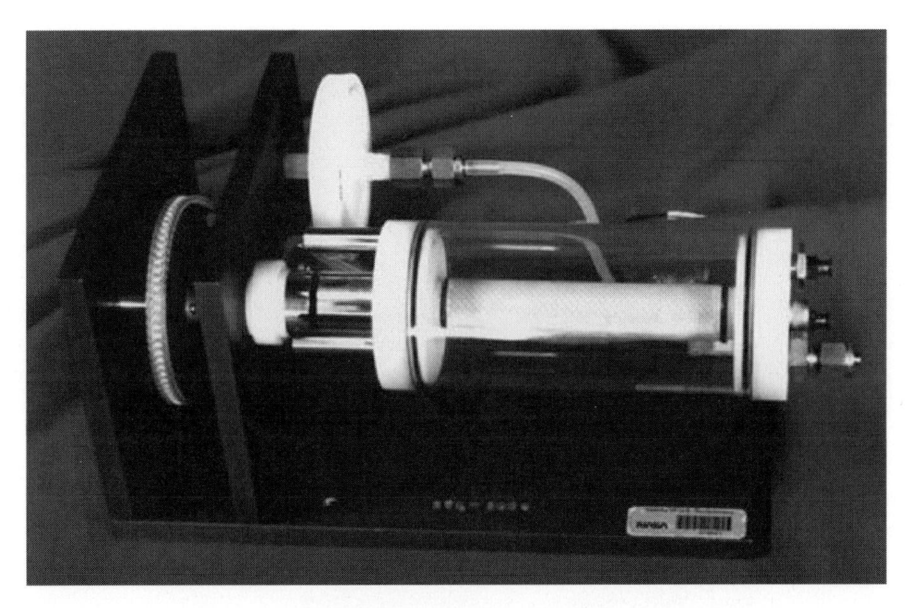

Figure 4.1. The bioreactor, developed at the NASA Johnson Space Center, is deceptively simply in appearance, but is revolutionising the study of healthy and diseased human tissue to understand the fundamental processes of cell growth and development. Future applications include the replacement of damaged tissue and, one day, the production of healthy organs for human transplants.

In trying to determine how to simulate microgravity for the cells being transported to and from space, Wolf reports: 'We took many wrong approaches, but finally, three of us came up with the approach to spin the culture, on the ground, in a cylinder of culture media.' But the researches were not only interested in protecting cells during travel. Similar to the way our blood services growing tissue *in vivo*, the idea, Wolf explained, was to 'build a machine that acted as a heart–lung machine, a kidney, a digestive system – essentially building a rudimentary artificial body to support the cells that were growing in this rotating, or spinning, cylinder.' The cells 'would be suspended in a fluid without introducing mixing devices inside the vessel which, by nature, disrupts the culture. We achieved a three-dimensional suspension of cells with a very quiescent fluid dynamic environment.' The three inventors of the bioreactor device were granted a patent in 1988 – one of 11 patents received by Dr Wolf.

The bioreactor is a cylinder that is rotated horizontally, and is completely filled with fluid culture medium. While a liquid, the fluid acts like a solid since there is no empty space in the cylinder, and rotates at the same rate as the cylinder, which results in a continuous suspension of the cells. The 'body' that supports the growing cells is a membrane covering the co-rotating cylinder centre post of the bioreactor, which allows for the efficient mass transfer of nutrients, and oxygen diffusion. The air from the chamber is pumped through a filter before it is returned to the bioreactor.

Figure 4.2. In the mid-1980s, specialists at the Johnson Space Center created the rotating wall vessel known as the 'bioreactor'. Here, technician Tinh Trinh, who conceived the design, checks hardware in preparation for its first test in space during the STS-44 Space Shuttle flight. The photograph was taken on 3 November 1988.

Dramatic advances in tissue culturing have been obtained using bioreactors or rotating wall vessels on the ground. In addition to the fact that the cells obtain a shape reflective of natural tissue growth, rather than flattened pancake shape on the bottom of a dish, they 'self-associate', to form a complex mix of collagens, proteins, fibres and other chemicals characteristic of *in vivo* tissue. This complex ordering of the cells into tissues, which is essential for the study of the effect of drugs, hormones, and genetic engineering, is a function of the connections between cells that normally form. This matrix material between the cells not only lets the cell know what kinds of cells are nearby, and how they should form, but also allows the tissue to respond to stimuli such as bacteria and wounds.

But it was recognised from the first rotating wall vessel experiments on Earth that this simulated microgravity also has limitations. Most noticeable is the fact that cell mass is limited to about a centimetre, or about 60 days of growth. Beyond that size the tissues reach a weight and density that makes them difficult to rotate in the cylinder in synchronisation with the fluid in which it is suspended, and once again they are exposed to the shear stresses that limit three-dimensional growth, and make contact with the surface of the vessel.

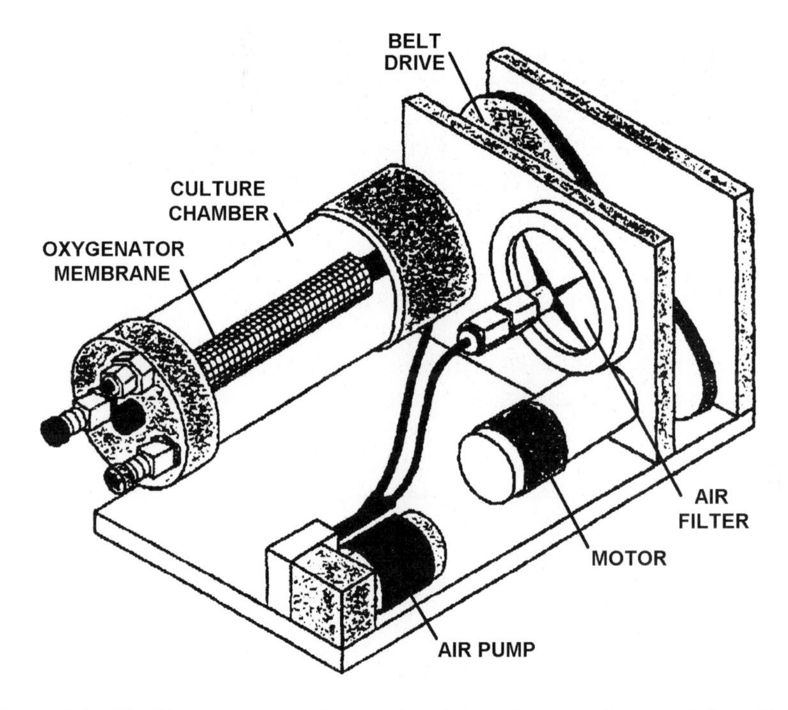

Figure 4.3. The bioreactor provides a reduced turbulence environment for cell growth, and allows the development of tissue constructs in three dimensions.

In other *in vitro* culturing techniques, when the tissue exceeds about one millimetre in size, there invariably develops a necrotic core, or dead cells, which are not being adequately serviced for nutrition and waste removal. And even in the ground-based rotating wall vessel, tissues have failed to produce natural blood vessel-like conduits for this purpose, which is one goal of experiments with the bioreactor in the microgravity of space.

Dr Pellis explained in testimony before Congress in 1997 that there are at least five critical stages in the engineering of tissue. The first is to produce an assembly of cells in three-dimensional arrays, after which the cells must propagate, or grow, in three dimensions. Synthesis of the appropriate intercellular 'cement' (matrix) to hold cells together and provide signals to the cells is necessary, and there must be a differentiation or specialisation among the cells for diversified functions. Finally there must be the formation of blood vessels to deliver nutrients and carry away waste.

On the ground the bioreactor must be rotated to simulate the microgravity that allows cells to remain in suspension. In space, objects float – not because there is an absence of gravity, but because the orbit of the spacecraft around the Earth means that the centrifugal force of its forward motion is in balance with the pull of the force of gravity toward the Earth; the spacecraft is constantly 'falling' towards the Earth. In the bioreactor the cells are also falling, but because of constant rotation they never go anywhere.

While it would be unnecessary to rotate the fluid mixture inside the bioreactor cylinder to keep the cells suspended, Dr Pellis explains that since the system developed for ground-based research – a rotating wall vessel – included not only the culture vessel but an entire integrated system to keep the cells alive, 'the same system was modified for use in space'. Adapting the existing system for experiments in space was far more efficient than specifically designing a device for space.

4.2 A NEW TOOL FOR GROWING TISSUE

In 1987, when the team at JSC had discovered how to culture cells in suspension, Dr J. Milburn Jessup was a cancer surgeon at the University of Texas M.D. Anderson Cancer Center. According to a 1998 article in the NASA Marshall Space Flight Center's publication *Microgravity News*, Dr Jessup's mentor there was I.J. Fidler, who was most interested in understanding the process of metastasis, or how cancerous cells spread from a primary to a secondary site in the body.

In testimony before Congress on 22 June 1993, Dr Jessup explained the motivation for his research. Colon cancer is the second leading cause of cancer deaths, while only three out of every 100 patients with pancreas cancer live for five years. 'The biggest problem with these cancers that we as medical scientists have been unable to solve is metastasis – the spreading of cancer. Cancer kills people young and old, not by simply growing into a large tumourous mass, but by invading other parts of the body, such as the liver, brain, or lungs. Once this invasion takes place it robs the patient of essential body functions until the patient dies... In the past,' he continued, 'new drugs were tested against the cancer cells grown on plastic or in mice, both poor representative of the human systems, since cancer cells grow and react in these conditions differently to drugs than do cancers growing in the patient. That is why our cancer treatments fail so often. The testing process for new drugs starts by looking at how well drugs kill cancer cells on plastic or in mice. But drugs that kill cells on plastic may not kill cancer cells in the patient.'

By early 1987 the bioreactor team at NASA had successfully grown cells and carried out two-cell co-culturing experiments in their new device, exploring the potential to develop three-dimensional use of tissue. They were ready to reach out to the broader scientific community to determine whether this promising new technique could be applied to their research. NASA researchers who had worked with Dr Jessup contacted him to see if growing malignant cells in three dimensions would shed light on this important feature. Jessup recalls that he was quite impressed with the rotating wall vessel system: 'This bioreactor prototype was likely to be rugged and durable.'

Jessup provided the team with simple human colorectal cancer cells, which were successfully grown in the fledgling bioreactor. Next, a co-culture experiment was attempted – in this case both normal and cancer cells – to determine what the interaction would be. Jessup described the results for *Microgravity News* in 1998: 'Fairly large tissue aggregates grew, and these had the ability to really recapitulate the morphology, or appearance, of what occurs *in vivo* in mice.' The experiment produced tissue masses resembling a three-dimensional cancer tumour, which had not been possible in other tissue culture systems.

In his congressional testimony, Jessup stated that during subsequent years of research, the rotating wall vessel developed at NASA 'has given us unique insights into the cancer cell than we would have had otherwise.' His team has shown that one colon cancer cell (MIP-101) produces a substance, CEA, which is produced by many colon, breast, stomach and other cancers, and bears on how the cancers metastasize. When cancer cells were grown in two dimensions on plastic or in mice, they did not produce the CEA as they do in the patient, but by using the rotating wall vessel, Jessup's group, and other cancer researchers, were able to grow these cells that clinically are most likely to kill the patient.

Jessup reported that his group has also found that cells cultured in the rotating wall vessel produce cell growth factors that are not produced in two dimensions. He believes that such factors are not triggered by the cells until they reach a high enough density to form tissue – up to about 1 centimeter, only possible in the rotating wall vessel bioreactor.

Even after Jessup's first successful experiment in the ground-based rotating wall vessel, many in the scientific community were somewhat skeptical of this radically new approach to solve an old problem. Dr Pellis told *Microgravity News* in 1998 that in the early 1990s a small group of researchers based at universities came to the JSC to use the new ground-based rotating wall vessel. As news of their results spread, NASA found that it could not meet the demand from researchers. NASA licensed Ray Schwartz and C.D. Anderson to manufacture the rotating wall vessel via Synthecon Inc, a new company that they had formed. By 1994 the devices were available on the commercial market. Today, more than 600 basic systems – which Synthecon calls the Rotary Cell Culture Systems – are in use internationally.

In 1994 Dr Pellis became director of NASA's tissue culturing programme at JSC, after 22 years of academic research in cancer immunology. At that time there were 14 investigations being sponsored by NASA. Today there are more than 125 investigators using the rotating wall vessel.

Dr Jessup had designed the first successful experiment with the ground-based rotating wall vessel, and provided the cells and the analysis of the results. He also provided cells for the first space-based experiment with the bioreactor in July 1995, as part of the STS-70 Space Shuttle mission. During the experiment the colon cancer cells aggregated to form masses 10 mm in diameter. Jessup reported that this was 30 times the volume of the cell samples grown in the ground control experiment. The crew aboard the Space Shuttle downlinked video images of the bioreactor tissue cultures during the mission. The video showed orange colon cancer cells coalescing in globules, some of which were described by Mission Specialist Mary Ellen Weber as being 'as large as a pea.' The same experiment was repeated in August 1997 on STS-85, and mature, differentiated cells were again produced. It was thereby confirmed that the microgravity of space affords an environment that significantly improves cell culture and tissue growth. A new tool in the study of cell biology was born.

Today Dr Jessup is at the University of Pittsburgh Medical Center, and as part of

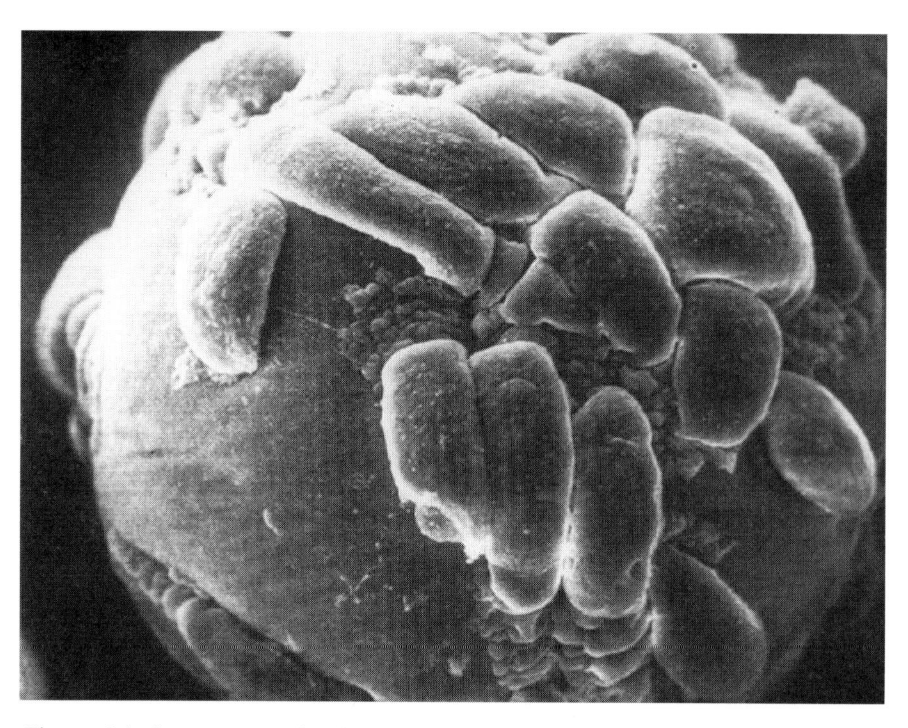

Figure 4.4. Some types of cell that are grown in the bioreactor are 'anchorage dependent' – they must attach to something to grow. Beads are used as the anchors in those cases – such as these cells that were grown in a ground-based rotating wall vessel – attached to a polymer microbead.

his continuing interest in space research he is the Chairman of the microgravity biotechnology discipline working group, which advises NASA. He has also conceived of intriguing roles for the bioreactor in space when astronauts are *en route* to planetary destinations. It can be used to culture red blood cells or skin in the event of an accident or trauma, he proposes, and can also be used to culture unicellular organisms such as blue-green algae, as a food supplement, or as a means of replenishing the oxygen for the air supply on the spacecraft.

Even though bioreactor experiments in space have produced results superior to the rotating wall vessel reactor used on the ground, opportunities to fly scientific experiments in space have been so limited (and expensive) that most of the research using this new bioreactor technique has been carried out with rotating wall vessels on the ground. Overall it has been found that the tissue cultured in three dimensions in this device is far superior to conventional two-dimensional tissue growth, and in many cases, animal models.

Some commercial enterprises involved in bioreactor research hope to be able to market new therapeutic products based on their experiments. Synthecon's C.D. Anderson reports that a commercial experiment that flew with former astronaut John Glenn on the STS-95 mission in the autumn of 1998 was successful in culturing a genetically modified protein which women secrete during pregnancy, and which prevents the abortion of the foetus. This protein also relieves the symptoms of arthritis. He is hopeful that approval from the Food and Drug Administration could be forthcoming in four or five years, after clinical trials.

4.3 SPACE TECHNOLOGY ON EARTH

By the mid-1990s it was clear that the rotating wall vessel technology that had been developed in the space programme could potentially revolutionise biology and medical research on the ground. The space agency launched an effort to involve the broader scientific community in this new frontier of research, and in 1994 NASA and the National Institutes of Health signed an agreement to establish a joint NASA/NIH Center for Three-Dimensional Tissue Culture at the Institute of Child Health and Human Development. NIH provided the laboratory, and NASA provides the rotating wall vessels and other support. Since the agreement was signed, more than a dozen laboratories within NIH have incorporated the bioreactor within their research capabilities. In 1998 there were 16 research projects underway.

At the joint centre there has been the successful culture of several infectious agents that are difficult to grow and control in a conventional culture setting. One of them is *cyclospora*, a parasite that lives in berries and causes extreme gastrointestinal distress when ingested. It is prevalent in developing countries. 'No one has been able to grow *cyclospora* in culture until this year, when researchers at the joint centre took a new approach and cultured the organism with cells from the small bowel', using a bioreactor, Dr Pellis stated in 1998. The investigators, Dr Darcy Hanes and Ben Tall from the Food and Drug Administration, observed that the parasites began to adhere to the human colon cells by the second day of culture growth, and that infection of the cells occurred. The tissue samples are being used to design therapeutic drugs or antibodies, and to design a strain of the organism from which a vaccine might be produced.

Researchers at the Center have also cultured human immunodeficiency virus (HIV-1). HIV had been grown before, but with the bioreactor, scientists were able to propagate the virus in human lymphoid tissue, to provide them with a picture of the full dynamic process of the disease. Researchers Leonid Margolis and Joshua Zimmerberg from NIH state that the 'critical events in HIV disease occur in human lymphoid tissue', which made this series of experiments critical to the understanding of the progression of the disease. They conclude that this tissue growth system can be used as a universal tool to study critical events in HIV infection.

Another team using the rotating wall vessel includes Dr Mike Bray of the United States Army Medical Research Institute of Infectious Diseases at Fort Detrick, Maryland. These researchers set out to demonstrate the ability of the ground-based bioreactor to act as a 'universal' pathogen culture system, for the primary isolation of previously unrecognised viruses, parasites and bacteria during outbreaks of newly emerging disease.

According to the Fiscal Year 1997 annual report of the Center, it took four months to identify the causative bacteria of the 1970s outbreak of Legionnaires

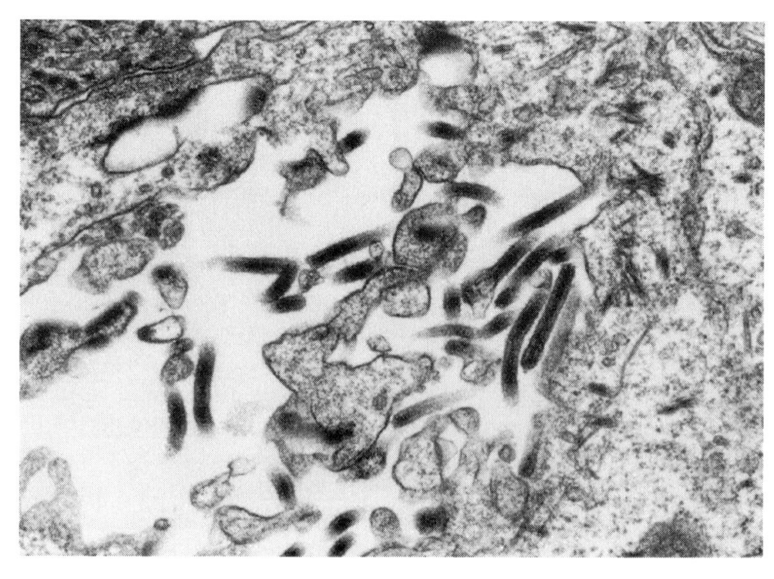

Figure 4.5. An increasing concern around the world is the growing incidence of newly emerging diseases, some of which have taken weeks to identify, using the standard tissue culturing techniques. Some of them, such as the *ebola* virus, have been succcssfully cultured in Synthecon's ground-based rotating wall vessel bioreactors. (Courtesy Institute of Infectious Diseases.)

Disease. The HIV-1 retrovirus required more than two years to be isolated, and it took seven weeks to grow the Hantavirus in culture that caused the 1993 outbreak of Myuro Canyon Disease in the United States. Considering the threat to public health from newly emerging diseases, the scientists were anxious to find a better method.

Known pathogenic agents which had proven difficult to isolate by normal protocols were introduced in the rotating wall vessel culture media along with human tonsil tissue. In eight days, *borrelia burgdorfei* was cultured in the tonsilar tissue, which is the agent responsible for the disabling Lyme Disease in the United States. The rotating wall vessel also demonstrated the ability to culture the deadly *ebola* virus, and the scientists plan to try to culture the causative agent of syphilis, which is 'notorious for its resistance to culture by conventional methods'.

The importance of being able to make timely determinations of disease-causing agents was demonstrated in the summer of 1999 in New York City where, on 23 August, health authorities were alerted to a possible epidemic involving neurologic symptoms, most probably carried by mosquitoes. In early September, after testing against only six insect-borne viruses already common in the United States, the Center for Disease Control (CDC) concluded that the outbreak was of Saint Louis encephalitis.

As the number of cases climbed, officials at a city zoo observed that their exotic birds, and crows in the zoo's vicinity, were dying of a strange disease. A scientist at the zoo contacted the US Army Medical Research Institute of Infectious Diseases, which confirmed that it was West Nile Valley fever, which is not endemic to the

United States, that was killing the birds. Finally, one month after the epidemic began, the CDC announced that it had made a mistake in identifying the causative agent for the outbreak.

Experiments by investigators Steven Hatfill and Paul Duray examined the causative role for HHV8 in the pathogenesis of Karposi's sarcoma. This form of cancer is observed to develop in patients with AIDS. Human spleen, tonsil, lymph node and skin tissue in a rotating wall vessel were co-cultured with cells harbouring the HHV8 viral genome. In four out of five experiments the cells were able to generate lesions in human spleen and lymph node tissue that were consistent with the clinical diagnosis of Karposi's Sarcoma. The investigators hope to better understand the molecular mechanisms by which this virus causes the cancerous transformation in tissue.

Other laboratories involved in the Center include the Naval Medical Institute, and the United States Army Medical Research Institute for Infectious Diseases. More than 40 of Synthecon's Rotary Cell Culture Systems are in use in laboratories of the National Institutes of Health.

Ground-based research, which has expanded to involve hundreds of scientists, has proven to be an invaluable tool in the study of disease, but there are numerous cases in which only the condition of microgravity itself can open up whole new areas in studies of human health.

4.4 FROM RATS TO GENES

Dr Timothy Hammond and his team, at the Tulane Environmental Astrobiology Center, have been carrying out research, jointly sponsored by Tulane University, Xavier University, and the VA Medical Center in New Orleans, on the protein receptors that bind common toxins in the proximal tubule, a microscopic tube in the kidney. The kidney, as it filters the body's waste products from metabolism, is often damaged by drugs and toxins contained in strong antibiotics. The brain and liver are also effected by environmental toxins.

In 1998 Dr Hammond explained to *Microgravity News*: 'We were interested in the proteins that get bound in the kidney by these toxins. We wanted to culture cells that make these proteins, to develop protective agents. But the problem is that there is no [conventional] cell culture line that expresses the relevant proteins... If you take a kidney cell and put it into a cell culture, a day later it no longer has any of its special features. So we had a lot of interest in finding a cell culture method that would keep the special features of tissues intact.' They wanted to retain the special features of differentiated cells, such as microvilli, which are hair-like structures found in some tissues. 'We tried the rotating wall vessel,' Hammond said, 'and to our shock, surprise, and delight, the tissue was beautiful. All the hair, the microvilli, grows on the cells, and they express all the specialised proteins we needed. The results were very dramatic.'

Hammond's team decided that the next step in the research would be to attempt an experiment in space, where the tissue culture would not be limited in size. 'If we were truly going to understand how different cells grow together to form a tissue with all its medical implications, we had to find some way to get out of the limits caused by gravity. That is why we wanted to fly the renal tubular cells.' His experiment flew with astronaut David Wolf on Mir from September 1997 to January 1998.

In an interview in 1997, Dr Pellis explained that David Wolf's series of tissue culture experiments were a 'risk mitigation' experiment to determine if cell culture modules could be used in space as a repository to provide a continuous supply of cells for the bioreactor. In a laboratory on the ground, he said, cells are frozen at –80° C, or kept in an incubator as a stock for experiments. In space, the goal is to keep cells propagating under the benefit of microgravity, until they are required for three-dimensional culturing. Wolf's job was to culture the cells in NASA's Biotechnology Specimen Temperature Controller – a cell incubator – and then transfer them to other modules, to rehearse the procedure for later transfer to a bioreactor. One goal of the experiment was to examine the population of cells after many replications, to determine whether microgravity exerts selective pressure on the replicating cells.

The scientists were particularly keen to determine whether the cells would be the same after perhaps 100 replications in microgravity, and whether there would be any pathology introduced in the cells over long periods of time in space, from either

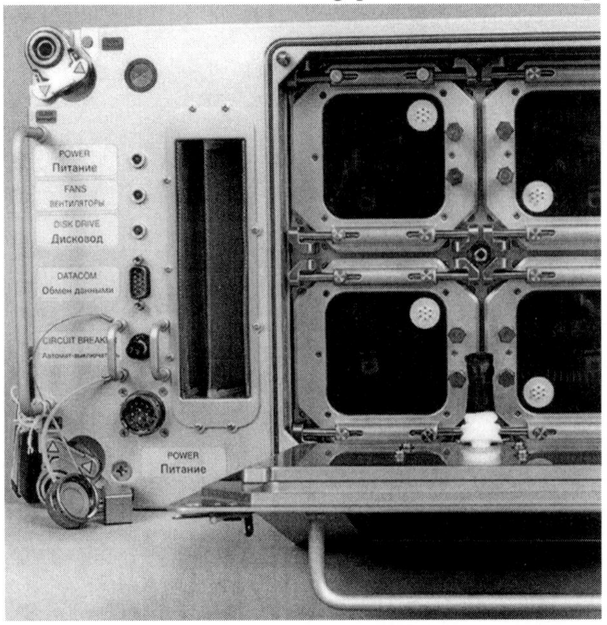

Figure 4.6. The Biotechnology Specimen Temperature Controller, operated by Dr David Wolf onboard Mir, is a device to cultivate cells until they are placed in the bioreactor. It is comprised of four incubation/refrigeration chambers, each holding three tissue chamber modules. The astronaut can monitor the controller through a colour computer display on the face of the facility.

Figure 4.7. During his stay onboard Mir, Dr David Wolf carefully noted the progress of Dr Timothy Hammond's rat renal tubular cells in the Biotechnology Specimen Temperature Controller, and of the microbial environment onboard the station, as seen here in the Priroda module.

weightlessness or radiation. Samples from cells produced at different times during Wolf's increment on Mir were archived, to determine whether there would be any 'drift' in the characteristics from the original cells. Dr Pellis already knew from his own experiments that immune system cells respond peculiarly in microgravity. The question was: does this happen to all cells in space?

One set of cells included in the Biotechnology Specimen Controller were leukemia tumour cells, which differentiate when changes are biochemically induced. Scientists were anxious to observe whether or not they spontaneously differentiate in space. A second set was composed of neuro-endocrine cells from the adrenal gland, which were used to determine whether they would produce substances under space conditions that normally provide signals for other central nervous system cells. Scientists hope space-grown cells may provide the life-like raw materials for nerve regeneration and the control of pain.

The third set of cells in David Wolf's experiment were Dr Hammond's rat renal tubular cells, which were a good match for Mir, because they would continue to grow and differentiate over the entire four months of the mission to help verify the long-term functioning of the hardware. Hammond reports that the tissue on Mir 'grew beautifully' under Wolf's care. 'We got gorgeous cell aggregates, bigger than the aggregates grown in the control experiment on the ground. And we saw proteins that we were interested in, the tubular toxin protein receptors, expressed in flight.'

In a first-hand description of the tissue culturing experiment on Mir during his stay there, Wolf told *The Scientist* writer A.J.S. Rayl that the rat tubular kidney cells 'grew in delicate, frond-like structures that were vastly different from conventional culture on the ground, and different from the rotating wall vessel.' He described the finished product as having a 'spectacularly different organisation... we got fluffy, large organising tissue, and, in fact, the whole bag turned into one piece of tissue, whereas on Earth, what you see are individual cells, or small clumps. You don't see them organising into one network.'

Although pleased with these results, Hammond and his co-investigators were interested in scheduling another space flight for their cells, to identify the mechanisms in the cell that *caused* differentiation and the expression of the proteins in microgravity. It was known that genes control these processes, but identifying the relevant genes out of the millions of genes in a cell seemed close to impossible. Hammond reasoned that a comparison of the genes that are active in the cells during culturing in spaceflight (where the tissue is more life-like, because more genes are being expressed) to those that are active in culture on the ground, which is less similar to *in vivo* tissue (where not as many genes are expressed) might help pinpoint the specific genes responsible for differentiation. Thomas Goodwin, a co-investigator from JSC, explained in an interview that to the extent that the ground-based rotating wall vessel 'emulates the conditions of microgravity', this should be seen as an intermediate step. The genes that change in the true microgravity environment are different, he said, and 'the effect is much more profound.'

In April 1998 the STS-90 Neurolab mission carried a sample of human renal tubular cells in the Bioreactor Demonstration System/Biotechnology Specimen Temperature Controller. This is not a rotating bioreactor, but a static tissue culture apparatus, which, in orbit, contains the cells, which remain suspended due to microgravity.

The experiment was designed to evaluate renal cells in their differentiation, or maturity of function, and their production of two hormones: erythropoitin (EPO), and an active variant of Vitamin D3. Renal cells were chosen because both EPO and Vitamin D3 are hormones produced by kidney cells, and are essential for normal health. They are two of the biggest selling pharmaceuticals in the world, and are prescribed for patients with kidney disease, AIDS, cancer chemotherapy and other immune function diseases. At present the American commercial market for these hormones is $2.5 billion per year. Therefore, any improvement in production of these hormones from space research, through a better understanding of the functions of renal cells, could advance the ability to treat an array of diseases worldwide. The scientists hope that production of stable, differentiated renal tissue may lead to implantable sources of these hormones.

In a 1999 paper that has been submitted for publication, Dr Timothy Hammond, Thomas Goodwin from JSC, astronaut Dr David Wolf and co-investigators explain that during the 1998 Neurolab Space Shuttle mission, the cells were cultured inside six bags incubated in the Biological Specimen Temperature Controller for six days, and then fixed by astronauts Rick Linnehan and Katherine Hire to preserve their characteristics, before being subjected to return from orbit. When the cells were returned to Earth they were analysed for the expression of 10,000 individual genes.

Figure 4.8. Following the successful culturing of rat renal tubules onboard Mir, Dr Hammond and collaborators flew a bioreactor experiment on the STS-90 Space Shuttle mission in April 1998, using a sample of human renal tubular cells, in order to study the activity of the genes that control differentiation in tissue. This photograph of Dr Hammond (left), James Kaysen (right) of the Tulane Environmental Astrobiology Center, and Thomas Goodwin (center) of the Johnson Space Center, was taken at the Kennedy Space Center as they readied their experiment for the Shuttle flight. (Courtesy VA Medical Center, Durham, North Carolina.)

The researchers found 'an amazingly large degree of change in gene expression at a steady state', indicating that the changes did not reverse. They report that 'a select but substantial group of more than 1,600 genes changed in each direction [up-regulated, or down-regulated, in terms of the magnitude of expression] in microgravity.' They defined a three-fold change as significant.

The genes whose expression changed the most are not surprising. They include adhesion molecules, cytoskeletal proteins, drug metabolising proteins, and other functions which are known to be altered in living systems that experience weightlessness. By comparison, only 914 genes were expressed when the same tissue was cultured in the rotating wall vessel device on the ground, indicating the diminished fidelity of simulated microgravity, compared to the real thing. Hardly any gene expression changes were observed in control tissue grown in two dimensions.

The scientists found that one transcription factor gene expression that underwent the greatest upward change was the Vitamin D receptor. They report that this finding 'extends our knowledge of the role of Vitamin D in tissue differentiation'.

This is important, because while it was 'initially identified for its role in calcium and phosphate metabolism, Vitamin D has recently been recognised to have broad influence on immunity, cancer surveillance, and the maintenance and development of tissue-specific cellular attributes.'

In an interview in August 1999, Hammond stated that the results from the experiment were 'beyond my wildest imagination'. They had 'no idea how big the genetic change would be or how many pathways there would be'. While they knew there was a minimum of shear and turbulence on the ground, no one knew the relationship between the physical changes and the biological expression. Their goal is to have a toxicology model, to have cells respond the same ways to toxins as do animals on the ground.

In discussing the difference in results between the rotating wall vessel used on the ground and the growth of cells in true microgravity, Hammond and his colleagues explain in their paper that as the tissue being grown becomes larger, shear forces develop in the vessel, and aggregates are torn apart. But in terms of the changes induced in gene expression, physical forces can induce changes, not through a physical sensor, 'but by the changes they induce in the culture environment.'

In an article on this research published *The Scientist* on 13 September 1999, Thomas Goodwin told author A.J.S. Rayl that the scientists' basic hypothesis is that

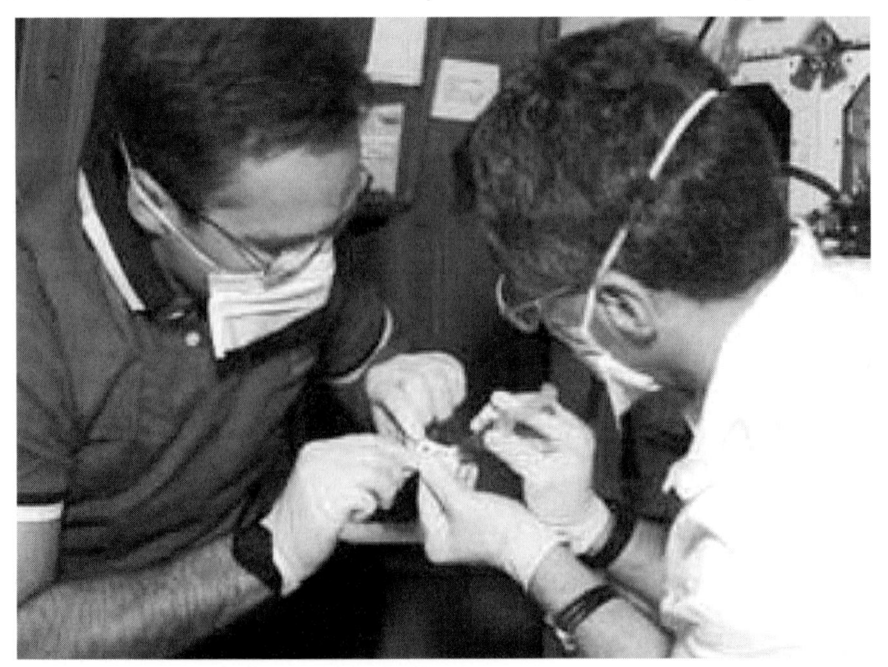

Figure 4.9. Before Dr Lisa Freed's bovine cartilage tissue was transferred from the Space Shuttle to the bioreactor onboard Mir, Shuttle astronauts Carl Walz and Jay Apt tested the growth media for the bioreactor for pH, and carbon dioxide and oxygen levels. They prepared the bioreactor experiment for transfer from the Shuttle to the Priroda module.

'gene expressions go through different phases as the embryo develops, and if you look at how the embryo actually develops – in a semi-weightless environment, a neutrally buoyant environment – it may be that a lot of the signals seen by the embryo may be recalled by cells when they're sent into microgravity.' While, Goodwin states, 'we have no substantial proof of that, it would seem from the anecdotal data we have so far that these cells respond in the way that they would respond if they were embryonic cells', whose job it would be to differentiate as they propagate.

Regarding the molecular genetic model possible to begin to construct from this initial experiment, Dr Pellis states that the gene analysis 'sets a benchmark that we can begin to use as a locator point, from which we'll launch a number of different investigations. It clearly shows that in the microgravity setting, the cells' response suite to microgravity is very unique and it's quite substantial. If you begin to look at what's re-regulated in the rotating wall vessel on the ground versus in space, you find that there is a congruent subset in both of the 1600 and the 800 genes. There are shared genes that are managed by the cell, and there are unique ones both in the ground-based bioreactor and in space. Like any model, it has a certain percentage parity with the real situation.'

As far as clinical applications are concerned, the researchers point out that the implantation of pancreatic islet cells grown in rotating wall vessels has been successful in diabetic insulin regulation for eighteen months, which leads them to believe that implants for other hormonal systems, such as EPO and Vitamin D from renal cells, will be rapidly developed.

In an interview on 23 August 1999, co-investigator Thomas Goodwin stated that the next step in this research will be to refly the experiment on the Space Shuttle to verify the results, and then, potentially, to fly a long-duration experiment on the International Space Station. In that environment, rather than being able to take the first sample after only a few days, they would attempt to examine at the transcriptional events as they occur in the first 24 hours. It is the transcriptional events that cause genomic expression shifts.

In the future, Goodwin proposes, the genomic expression in cells could be purposively modulated, and used to alter the progression of growth and differentiation. If the cells grown in space are more 'plastic', he explains, and they continue to grow, it would be possible to create organs as a consequence of knowing which genomic expressions are associated with growth.

As early as 1993, Professor Larry McIntire, of Rice University, proposed that if the manipulation of genetic materials or cell types can be preserved once the cells are back on Earth, 'the very high cost of manipulations in space would be highly leveraged, as actual manufacturing of pharmaceutical products would be Earth-based.'

There is also the promise of being able to develop genetic manipulation, or gene therapy, by studying the effects first in microgravity to produce the best results. Every gene, Goodwin explains, has some potential for expression. Some 'shut down for life' after the initial development phase. Other tissue, such as part of the colon, can regenerate. But every cell has a different gene expression, based on the activities

Figure 4.10. Astronaut John Blaha operates the bioreactor during his increment on Mir.

it carries out. If scientists can discover the mechanisms by which genes are 'turned off and on', it may be possible to 'turn on' cells as if they were still in their embryonic stage, and regenerate tissue, which is one of the promises of tissue engineering.

4.5 HELPING THE WAR AGAINST DISEASE

In 1993, in testimony before the House Committee on Science on the 'Health Benefits of Space Station Research', Dr Neal Pellis – then Associate Professor in the Departments of Surgical Oncology and Immunology at the University of Texas M.D. Anderson Cancer Center, Houston – discussed the research he had been conducting into one of the human body's most critical functions. He explained that we rely on our immune system to fight infectious disease and cancer. Immune cells must migrate through the body's tissues to the afflicted sites to attack any foreign invader. To do this, they must be able to 'crawl' through the matrix of material that binds the cells together. This matrix is comprised of many proteins, and includes the 'glue protein' collagen, which is a major constituent.

In the medical community, Pellis stated, to fight disease we seek ways to 'promote tissue invasion by immune cells called lymphocytes.' On the other hand, medical practitioners seek to minimise an immune response to protect transplanted organs from rejection by the body. And in auto-immune diseases, the immune system attacks a healthy organ or tissue. Auto-immune diseases, such as rheumatoid arthritis and lupus erythematosus, occur much more frequently in women, and female reproductive functions can even be effected, as in endometriosis. In all three of these cases, Pellis pointed out, 'it is clinically advantageous to be able to control the movement of lymphocytes in tissue.'

Researchers at the M.D. Anderson Cancer Center, Pellis reported, had discovered that as the immune system cells approach the site of growing cancer cells, they cease movement and fail to reach and destroy the diseased tissue. Experiments on immune system cells using the rotating wall vessel revealed a significant impairment in their ability to migrate through the collagen matrix that holds the cells together. Researchers had believed that this characteristic was unique to cancer, but studies of the response of immune system cells in microgravity, using the bioreactor in space, revealed a similar problem of immune system cells being unable to 'walk' well in space. Until Pellis' experiment on Space Shuttle flight STS-54 in January 1993, scientists knew that there was an impairment of the immune response of astronauts, but they did not understand the nature of the mechanism that produced the response.

In addition to providing this new window on the cause of the diminution of immune response in microgravity, researchers have continued to carry out biochemical analyses of the molecules that are involved in the attachment of immune cells to the matrix material and how that effects their locomotion. They are also examining changes in gene expression in microgravity, and the effects of microgravity on the matrix itself.

Dr Pellis stressed that the ground-based bioreactor was also an important capability to test 'agents that promote the invasion of tissue by immune cells' as a first step towards identifying possible treatments. When such agents are identified, he said, they may then be entered into preclinical trials in animal model systems.

4.6. GROWING SPARE PARTS USING SPACE TECHNOLOGY

During testimony before Congress in 1997, Neal Pellis discussed the possible tremendous impact on health of engineering human tissue using bioreactor technology. The annual cost of treating patients that suffer tissue or organ loss from diseases and accidents annually exceeds $400 billion, he stated. At present, the only treatments are 'maintenance strategies' such as insulin for diabetics and dialysis for kidney patients, or transplant of tissues and organs, which are limited by the shortage of donors and enormous cost.

However, imagine not only the savings in cost, but the reduction of human suffering and lost productivity, if whole new body parts could be 'manufactured' and implanted into those in need. In the short term, both ground- and space-based bioreactors will be used as laboratories to study the fundamental biological processes of life and disease. But it is possible that one day, not too far in the future, tissue that is grown in three dimensions from healthy cells will be used directly to benefit patients on Earth.

The project for transplantation of healthy tissue that is the closest to fruition is the effort to culture human pancreatic islet cells for diabetic patients, to control insulin production. About half of the blindness, kidney failure and limb loss in the

United States is due to long-term complications of diabetes, for a public cost of $1 out of $7 spent on health care.

In the early 1990s Dr David Scharp, Professor of Surgery at the Washington University School of Medicine in Saint Louis, Missouri, began studies of culturing human islet cells in a rotating wall vessel. He was most interested in the clinical applications of this new technology, and the possibility of pancreatic islet transplantation for diabetic patients, to produce insulin.

Although his studies with this new technology were just beginning, he was optimistic that the NASA-developed bioreactor device would reduce the net loss of islet tissue, which is up to 50% when using conventional techniques, due to the steps taken to reduce the immune reaction of the host patient, to prevent rejection of the new tissue. His optimism was well founded.

In 1997, NASA and the Juvenile Diabetes Foundation signed a Space Act Agreement for collaborative work, including research using the rotating wall vessel bioreactor, to understand how to cultivate and transplant Beta cells into Type 1 diabetics. Based on the success of the ground-based experiments, a Technology Transfer Agreement was signed with the VivoRx company in Santa Monica, California, to develop a treated seaweed membrane encapsulation method for the implantation in the abdomen of cultured human islet cells. The seaweed allows insulin and glucose to diffuse back and forth so that the transplanted cells work as an artificial pancreas, which allows them to be accepted by the diabetic patient's immune system. Once transplanted, the cells secrete the appropriate amount of insulin for regulating the body's blood sugar levels. The results in human volunteers have been highly promising, and the technique is now undergoing the third level of testing by the Food and Drug Administration, toward approval for widespread use.

Another research project that is aimed at replacing damaged or diseased tissue is being led by Dr Lisa Freed, at the Massachusetts Institute of Technology. Her aim is to attempt to produce masses of cartilage large enough for implant studies. Cartilage serves as a basic structural material in the human body, such as in the nose and ears, mainly consisting of cells known as chondrocytes. It is also a connective tissue, and is characterised by its nonvascularity (lack of blood vessels). The goal of the cartilage experiment that Dr Freed designed for the Mir space station was to investigate the patterns formed by the cells as they attach into clusters, and the interactions that lead to the production of functional tissue.

The STS-79 Space Shuttle mission that delivered astronaut John Blaha to Mir in September 1996 also delivered Dr Freed's cartilage cell cultures to be grown for the first long-duration test run for a bioreactor. The bioreactor was housed in the Biotechnology System facility that can service a number of experiments. It had earlier been brought up to the Mir station inside the Priroda module. The cells had been cultured on the ground on synthetic polymer scaffolds, which induces cell differentiation and degrades at a defined rate, for three months in a rotating wall vessel, and then cultured on Mir for four months.

Marshall Space Flight Center's *Microgravity News* published a real-time report on Dr Freed's experiment in its Winter 1996 issue, written while the experiment was in progress. Thomas Goodwin from JSC, who was a co-investigator for the experiment,

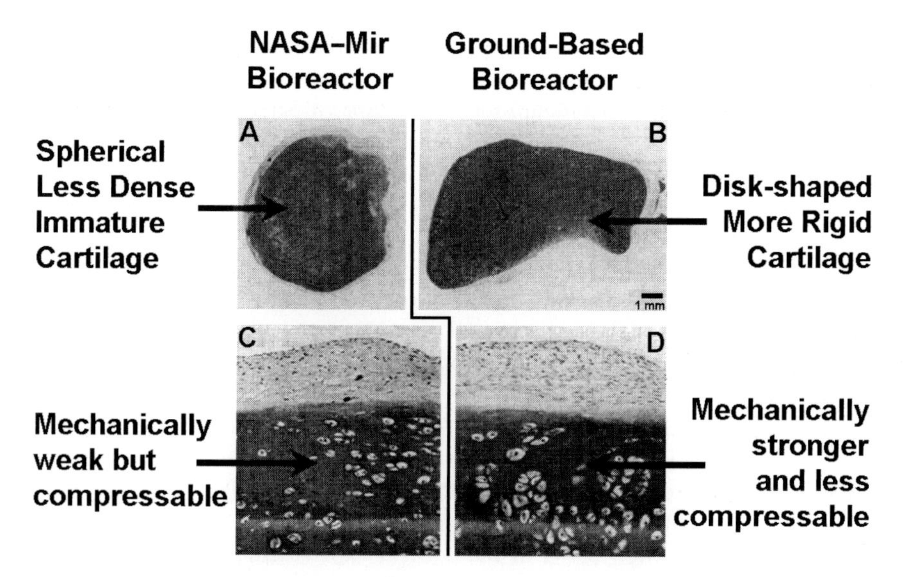

Figure 4.11. Dr Freed's bovine cartilage experiment in the bioreactor onboard Mir produced tissue that was less mature and more mechanically weak than that grown in the rotating wall vessel bioreactor on the ground. This less mature tissue may be important for tissue transplants in the future. Dr Freed found that both space-based and ground-based bioreactor products were superior to those produced with standard culture techniques.

explained that cartilage was chosen because of the potential benefit from the extended exposure to microgravity, in the hope of cultivating masses of tissue large enough for implantation studies.

Dr Pellis also stressed that bovine cartilage cells were chosen because they are so hardy. He explained that any ordinary cut of beef at any meat counter contains cartilage, which is the white fibrous tissue between the ribs. He pointed out that these cartilage cells are still alive when purchased at the store, typically two weeks after slaughter, and that 'its durability was right for this kind of adventure.'

At the time of the cartilage experiment onboard Mir, Principal Investigator Freed had experienced success culturing cartilage tissue in the rotating wall vessel, during a Space Shuttle mid-deck experiment. But her Mir experiment was the first long-term trial of the bioreactor, which posed several challenges for the design and operation of the equipment. The team at JSC, according to *Microgravity News*, carried out sophisticated and extensive testing of the computer to control the bioreactor's conditions, and selected the exceedingly hardy cartilage cells to grow in what was then an untried situation. This was a wise decision.

On just the second day after the bioreactor was transferred from the Space Shuttle to the space station, Mir astronaut John Blaha reported that experiment parameters in the unit did not match those of the control experiments on the ground. Without a match, no direct comparison would later be possible, at the completion of the experiment. It appeared that the problem lay with the control computer.

Luckily, the Space Shuttle was still docked to the Russian space station. Although the Russians had to enforce very restricted communication with the Mir crew due to their very limited ground-based communication capabilities, the Space Shuttle, using a constellation of communications satellites, is in near-constant contact with Mission Control in Houston. The Russian flight director in Moscow Mission Control circumvented regulations and allowed the Shuttle's communication links to be used to talk directly with Blaha onboard Mir.

To troubleshoot the problem, the JSC ground engineering team, led by Rafael Garcia, asked Shuttle astronaut Jay Apt to photograph both the experiment's exterior and the cells inside the bioreactor, using an electronic still camera on the Shuttle. Although Apt was able to photograph only the outside of the bioreactor unit because the interior light was not working, it was revealed from the images that a pin cable connecting the bioreactor to the computer was not properly locked in place. Garcia's ground team decided to turn the experiment off for a brief period of time, to allow Blaha to check and reconnect the cable, which solved the problem.

In late October, *Microgravity News* reported, ten cartilage samples seeded in the bioreactor were larger than those in the control experiment units. Astronaut John Blaha also reported that spin-off pieces of cartilage had started to grow in the bioreactor. Thomas Goodwin explained that this indicates that the outer cells of the original ten tissue samples are still growing and dividing, and that some of these cells drifted away and attached to other drifting cells to form new aggregates of tissue. Scientists had expected that this might happen, but were surprised it had shown this behaviour so soon in the experiment.

'The remarkable thing about the new foci [spin-off sites]', Goodwin told *Microgravity News*, 'is that chondrocytes, or cartilage tissues, grow very slowly. In preliminary ground-based tests, we didn't see this phenomenon until the very end of a 125-day experiment, and then it was very limited – maybe two or three foci. In the Mir experiment, we are seeing six to eight foci and we're only 50 days into the experiment. We aren't seeing anything like this in the two control experiments.'

The December 1997 issue of the *Proceedings of the National Academy of Sciences* contains a detailed analysis of the results of Dr Freed's Mir experiment. The cartilage tissue grew from September 1996, and was brought back to Earth onboard the Space Shuttle at the end of mission STS-81 on 22 January 1997.

The scientists report that bubbles were observed in the Mir bioreactor between flight days 40 and 130. As seen from the video footage taken by Blaha, the bubbles did not come into direct contact with the tissue. A preliminary assessment of the Mir tissue was carried out only 30 hours after landing, and were comparable to Earth-grown control samples, in that 95–99% of the cells were viable. But the tissue grown in space 'tended to become more spherical, where those grown on Earth maintained their initial discoid shape', the scientists reported, because in microgravity, where the cells floated freely they grew equally in all directions. The space-cultured cells were also smaller than those on the ground.

The most striking and critical difference between the Earth- and space-grown cartilage was that the 'mechanical properties of the Mir-grown constructs were inferior to those of the Earth-grown constructs'. This result was not altogether surprising, because it is well known that musculoskeletal tissue, such as bone and cartilage, 'remodel in response to physical forces, and are adversely affected by spaceflight'. The loss of bone and muscle mass has been observed in every space traveller.

Discussing Dr Freed's results in an interview in 1997, Dr Pellis reported that while the ground-based tissue grew within the boundaries of the scaffold, and the tissue in the bioreactor exceeded the boundaries of the scaffold by at least 50%, the mass of the space cartilage was only one third that of the ground control cartilage. He attributes this to the fact that the natural development of cartilage tissue is related to the load that is placed on it, similar to bone, and that in space the tissue is more compressible and less stiff because of the lack of load, or stress, without the weight resulting from gravity. Indeed, Dr Freed's cartilage tissue seemed to be suffering from the same microgravity effects suffered by astronauts and cosmonauts. The results, therefore, provide an important avenue for studying the causes of this debilitating effect of adaptation to microgravity.

Dr Pellis also points out that this 'immature', less dense cartilage tissue grown in space may be found to be well suited for transplantation, because studies indicate that it forms a better bond than mature tissue, with already existing cartilage. The space-grown tissue is most like embryonic cartilage tissue, he explains, and it may be more 'biocompatible' for repair of cartilage, which may also have implications for the repair of bone or damage from wounds.

Dr Freed and her colleagues have been involved in another application of rotating wall vessel bioreactor research, with significant potential applications for the replacement of diseased or damaged tissue, in the human heart. NASA's *Space Science News* reported in a release on 5 October 1999 that Dr Freed and her co-investigators have published their findings. In 1994 the research team completed the first successful experiment in engineering heart tissue.

Figure 4.12. The Synthecon Rotary Cell Culture System is being used by hundreds of researchers around the world to increase the fidelity of tissue cultured in Earth-based laboratories. This rotating wall vessel system is being used in cancer, HIV and other disease research, as well as genetic studies, tissue regeneration and drug screening.

Cells that had been 'seeded' on a three-dimensional scaffold, to provide a structure on which they would grow inside the rotating wall vessel, actually began beating as one piece of tissue, outside a living body. 'No one had ever done this before', said Dr Freed. The current research papers provide an analysis and description of the structural and electrical properties of cardiac tissue. Heart function depends upon the ability of the tissue to conduct electrical impulses. The researchers compare how various parameters differ in different cell culture environments, in order to optimise the conditions for cardiac tissue engineering.

Dr Freed, her collaborator at MIT, Dr Gordana Vunjak-Novakovic, and the other members of their team are continuing their research. They hope to eventually be able to use engineered cardiac tissue to repair damaged heart tissue, test new drugs, and perhaps grow replacement hearts. While that is far off in the future, Dr Freed states that they have 'taken the first step'.

Scientists in laboratories around the United States are at work to solve the problems of growing difficult liver, kidney, and other tissue in rotating wall vessels, to eventually produce entire organs. Dr Wolf is also hopeful that neural regeneration will be possible through tissue engineering. 'I have a friend who had a broken neck', he stated in an interview, '[who is] one of the best pilots I've ever known, and there are a lot of people like him we need to help. We had nerve cells on my mission, and they showed an excellent ability to reproduce in zero gravity'. Dr Wolf optimistically states: 'There's no question that one day humans will grow real organs for reimplantation.'

4.7 ADVANCES IN WOMEN'S HEALTH

In 1990, statistics from the National Cancer Institute indicated that the direct medical costs in the United States alone due to cancer were $35 billion, accounting for 10% of the total cost of disease in the US. The American Cancer Society now projects that by age 85, one in every 65 American women will develop ovarian cancer, and one in every nine women will be diagnosed with breast cancer, which is the second major cause of cancer deaths in women.

In testimony presented on 22 June 1993 before the House Committee on Science, Space, and Technology, Subcommittee on Space, Dr Jeanne Becker, Assistant Professor at the University of South Florida Department of Obstetrics and Gynecology, explained that the goal of her research is to develop improved models for studying human cancer. She was attracted to the possibilities of growing cancer tissue in rotating wall vessel systems, she said, because of the ineffectiveness of using standard culture techniques. This was demonstrated by the fact that from 27 breast cancer patients she was able to grow only five specimens large enough to fill a dish.

In Dr Becker's experiments, tumour cells are suspended on beads in a liquid culture medium. The cells grow in nearby aggregates and form cellular bridges, enlarging the three-dimensional structure. Using the ground-based rotating wall vessel, she has produced clusters of ovarian and breast tumour cells 'which exhibit remarkable similarity to tumours grown within the body'. These types of tumour have proven difficult to grow in conventional tissue culture.

Reporting to the members of Congress on the experiments she had conducted up to 1993, Dr Becker said that ovarian cancer cells she cultured were taken from a cell line which was particularly unique, because it was derived from an ovarian tumour which consisted of several kinds of malignant cell. When they isolated a single type of cell present in the specimen, and cultured it under traditional flat, monolayer conditions, only a single cell type is seen, with all the cells showing the same appearance as is characteristic for one cell line. But when the cells were grown in the rotating wall vessel for 32 days, 'we were able to recover the multiple malignant cell populations, as were present in the original tumour... We concluded from these experiments that these ovarian tumour cells represent a type of stem cell capable of growing into many different kinds of cell.' The conditions in the rotating wall vessel 'allowed these ovarian tumour cells to differentiate into the multiple types of malignant cells as populated the tumour originally.'

Breast cancer cells were cultured directly from the patient specimen. Dr Becker has successfully cultured a tumour from five different patients, and all tumour specimens showed three-dimensional cellular growth. They grew very slowly, requiring up to 90 days in the vessel to achieve significant growth. 'The inability to adequately culture primary breast tissue has been a limitation to establishing better models to study this type of cancer', she explained.

In 1993 Dr Becker reported that the next planned step was to test the sensitivity of the tumour cell aggregates to 'various forms of biologic and hormonal therapy and to chemotherapeutic agents.' New forms of anti-tumour treatment can also be tested. They will also examine alterations in cancer-associated genes. 'Our goals for the rotating wall vessel models include kinetic studies of growth rates and cellular divisions occurring during culture in the vessel, so alterations in these parameters can be examined at defined time points during three dimensional development,' she further elaborated in a 1994 article. 'The models will be instrumental in studying alterations in the cancer-associated genes that occur during tumour progression. By examining the cellular genome, this line of investigation addresses the very basis of why a tumour cell is a tumour cell. The ability to better study the genetic make-up of a cell as it grows in the body could lead to an improved understanding of how these tumours metastasise.'

Dr Becker has continued her cancer research with the rotating wall vessel, and in a May 1997 paper, reiterated that 'although animal models are a successful mode for propagation of human tumours and are useful under certain experimental conditions, many complex biochemical and molecular studies require the propagation of cells in the absence of confounding host responses.' In recent experiments, extensive analyses were carried out on the tumour tissue grown in simulated microgravity. Immunocytochemical analysis, scanning electron microscope analysis, flow cytometric analysis of proteins and cell cycle analysis were performed, and proto-oncogene expression was determined.

The immunocytochemical analysis revealed the tissue more closely resembled the strong reactivity of the original tumour than conventional tissue culturing. The

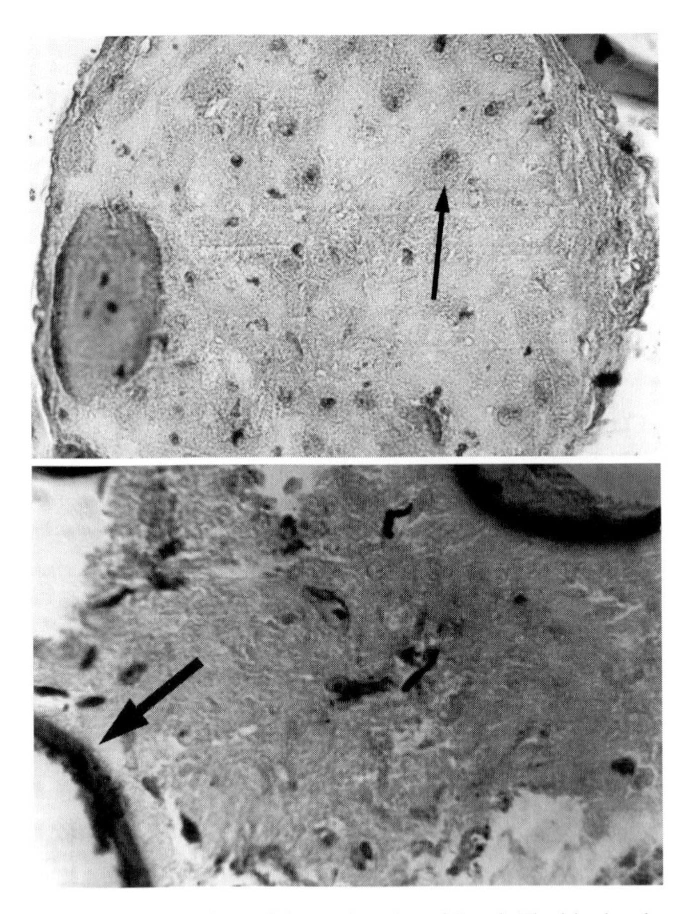

Figure 4.13. Dr Jeanne Becker, of the University of South Florida, has been engaged in intensive studies of the cancers that are prevalent in women – breast and ovarian cancer. The upper photograph shows tissue grown after 56 days in a rotating wall vessel bioreactor on the ground. It reveals areas of tumour cells dispersed throughout the cell background. The arrow denotes the focus of the breast cancer cells. The lower photograph is a higher magnification view of the same tissue. The arrow points to the surface of an anchor bead with breast cancer cells. (Courtesy Dr Jeanne Becker.)

scanning electron microscopy demonstrated the ability of the cells to produce secretions and differentiated surface structure. It was found that there was expression of RNA for cellular (cancer-causing) oncogenes and the p53 tumour suppressor gene. The conclusion of the scientists was that 'culture in the rotating wall vessel promotes the proliferation of... cells in three dimensions, allowing the formation of an organised tumour-like architecture.'

During a Symposium on 21 May 1998, sponsored by George Washington University in Washington DC, and titled 'NASA Research and Human Health', Becker reported that she is now studying how and why drug resistance develops against cancer, which is the main problem with chemotherapy. She is also measuring

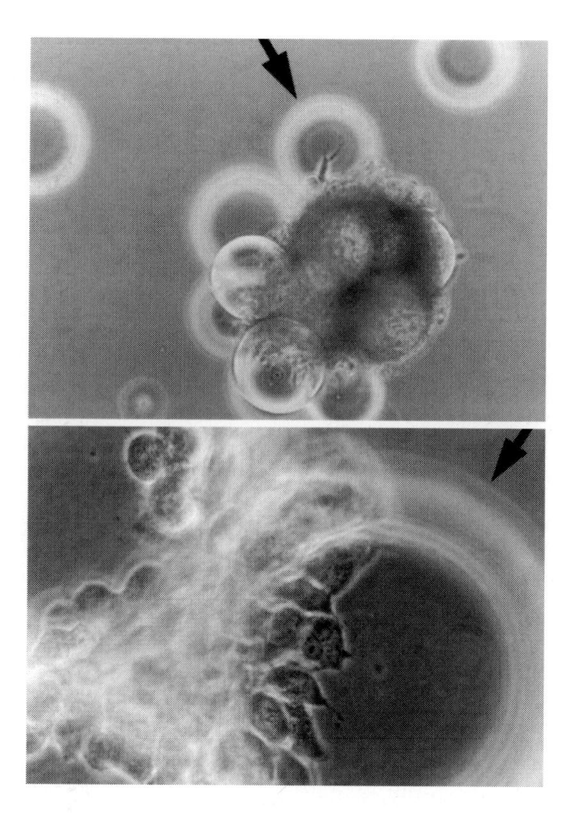

Figure 4.14. These images from Dr Becker's research show breast cancer cells aggregated on micro-carrier beads. The lower photograph is at a higher magnification, illustrating the intercellular boundaries and the important achievement of three-dimensional tissue outward from the bead. (Courtesy Dr Jeanne Becker.)

the hormone responsiveness of tumours as they grow in the rotating wall vessel. Her hope is that new and more effective treatments will emerge from what can be learned about the onset and progression of cancer in women.

In addition to bioreactor technology's contribution in discovering new treatments for breast cancer, it can also aid in the understanding of how healthy cells become carcinogenic in women. NASA's Space Science Report from the Marshall Space Flight Center, dated 1 October 1998, reviews the mammary cell studies being conducted at NASA's Huntsville, Alabama Center. Dr Robert Richmond is the Director of the new Radiation and Cell Biology Laboratory at MSFC, and is also a research Associate Professor of Medicine at Dartmouth Medical School in Hanover, New Hampshire. He is culturing non-cancerous, healthy cells, 'hoping to learn what guides their growth, and how we might use that knowledge to thwart malignancies before they are created... The type of mammary cells we are growing comes from an individual [genetically] susceptible to breast cancer,' he explained, 'and that susceptibility is likely driven by damage caused by ionising radiation.' The research serves a two-fold purpose. 'Space exploration will involve slightly increased exposures of crew-members to radiation, so what we learn from these cells could help justify methods of female crew selection.' This research could also 'help manage breast cancer in the national population at the same time.'

Dr Richmond is developing a breast tissue-equivalent model, which is a scientific description of how healthy breast tissue grows. He is using tissue donated by a healthy young woman who carries a defective gene which increases her risk of breast cancer five-fold. She elected to have a double mastectomy to reduce her risk of breast cancer, and donated the breast tissue for scientific research. This, in addition to Dr Becker's work with malignant cells, allows comparative studies of healthy and diseased tissue, which will increase the efficiency of testing new realistic therapies.

Hormone therapy, for example, is an option for breast cancer patients, because it is well known that breast tissue responds to estrogens. However, two-dimensional laboratory breast tissue does not demonstrate any response to estrogen. Dr Richmond is producing three-dimensional breast tissue constructs in the rotating wall vessel, and if this more high-fidelity tissue responds to estrogen, hormonal therapies that more closely match the desired response can be developed, with fewer side effects for the patient.

4.8 REPRODUCING COMPLEX TUMOURS

Tissue in the human body is interlaced with blood vessels that deliver nutrition to, and remove wastes from, our organs. When cancerous tumours form, they trigger the creation of a blood supply to help feed their abnormal growth. Without this vascularisation, tumours would not be able to grow without limit. Therefore, to continue to increase the fidelity of cultured tissue it will be necessary to grow tissue that can develop its own blood supply system. This is a necessary next step in cancer research, and was the object of the bioreactor experiment carried out on the very last increment of an astronaut on the Mir station – that of Dr Andy Thomas in 1998.

The purpose of the experiment was to co-culture breast cancer cells with cells to form blood vessels. This was the first attempt at vascularisation of a solid tumour in microgravity. In an interview conducted while Andy Thomas was onboard Mir with the bioreactor, co-inventor, medical doctor, and astronaut David Wolf said: 'We've been looking forward to this experiment for quite a while. Co-cult simply means growing more than one cell type at the same time, together. We've taken this to a reasonably advanced level on the ground in a simulated microgravity environment... The two types of cells Andy is growing are breast cancer cells, and a fibroblast layer, made up of angiogenic cells, that is a blood vessel-forming type of cell. These are two types of cell whose relationship is very critical. A tumour attracts blood vessels just to feed its cells, and this is a key area of cancer research – to find out why these tumours attract their own food supply.'

'One goal', Dr Wolf continued, 'is to reproduce breast cancer tumours in a fashion that even more accurately represents how these tumours grow in the body. It is not necessary for actual vessels to form; it is the interaction between the two cell types that is of great importance. We chose this pair of cells because it takes us a

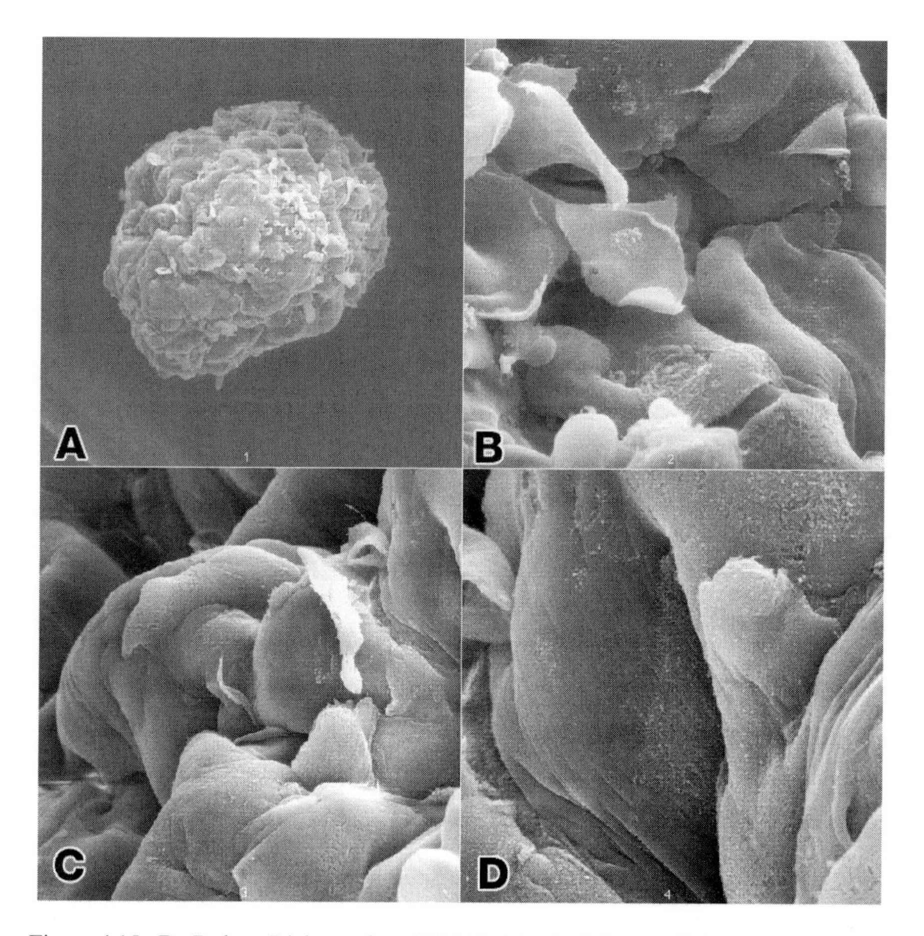

Figure 4.15. Dr Robert Richmond, at NASA's Marshall Space Flight Center, has been studying healthy tissue from a cell line that is susceptible to cancer. In these images, human mammary epithelial cells from tissue susceptible to cancer is cultured with fibrous tissue from the same initial breast tissue, to more accurately mimic the *in vivo* growth. Although this coculture of two different types of cell produced smaller tissue samples than when only the epithelial cells were cultured, Dr Richmond reports that the cell-to-cell interactions are different than when only one cell type is grown in the rotating wall vessel bioreactor. (Courtesy Dr Robert Richmond/Marshall Space Flight Center.)

good way down the path of studying breast cancer tumours, as well as helping us take the next step toward organ or tissue engineering', if the attempt at vascularisation is successful.

'It's very clear', Dr Wolf stressed, 'that in the end, there is no question that we will one day grow replacement organs for people, and space will unlock many of the secrets of how to do that. This is our first step in space – to move toward vascularising tissue.'

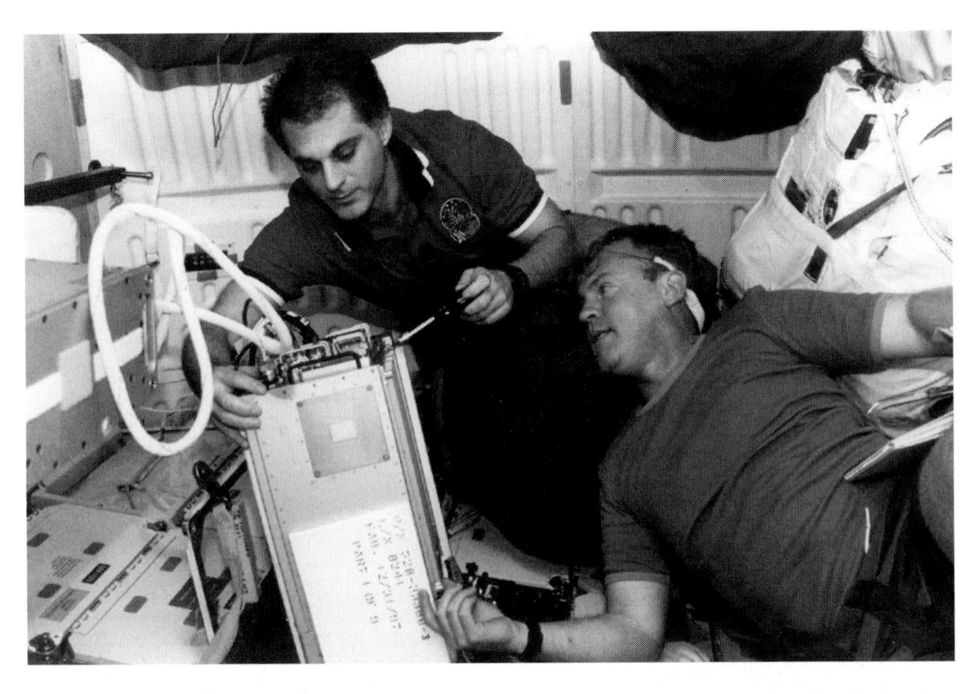

Figure 4.16. The bioreactor experiment onboard Mir during astronaut Andy Thomas' increment, the last on the Russian station, was designed to grow the first co-culture, of two different types of cells, in space. In this photograph taken during the STS-89 mission in January 1998, Mir astronaut Dr David Wolf (left), one of the inventors of the bioreactor, checks the bioreactor hardware, accompanied by Dr Andy Thomas, who replaced Dr Wolf on Mir.

I asked Dr Wolf if this could not also work in reverse; if we can discover what the mechanisms are that cause cancer tissue to create its own blood delivery system, could we not treat cancer by preventing the malignant tissue from being able to do that? 'This would be wonderful', he replied, 'once we understand the mechanism, the process, that gives us a directed target to be able to interfere with the process – this is one important strategy in fighting tumour growth.'

In response to a question concerning why this particular type of cancer cell was chosen for the co-culture experiment on Mir, Dr Wolf replied: 'We have studied upwards of 15 different types on the ground. We felt that breast cancer was a good model. It grows in a glandular formation on the ground. We just had to pick one, at some point. Colon cancer is another one we are ready to study in space. We just haven't had a chance yet. In the end, we will study all of them... You have to realise that we've had extremely limited time in the laboratory in space. We're trying to gather a database to help us direct and focus our research on the International Space Station. At that point, we'll have more extensive facilities and continuous time for many years in the laboratory. This type of work takes a whole team of researchers many years in the laboratory, and what we have [on Mir] is a small area of a small space station, with one astronaut, part time. We're really in our infancy.'

In an interview with Thomas Goodwin, one of the investigators of the experiment, a year and half later in July 1999, he reported that there are indications in the tissue that came back from space that vascularisation was beginning, and that the scientists are 'cautiously optimistic' about the results. Dr Pellis concurred, in an interview on 4 August, that the analysis of the tissue is 'long and arduous', and not yet complete.

Unfortunately, some of the hardware on Mir interfered with the growth of the co-culture tissue. Watching the growth inside the bioreactor, Andy Thomas noticed that there were bubbles forming inside. In a NASA Mir mission report of 13 March 1998, while the mission was in progress, Dr Neal Pellis explained: 'The bubbles are a problem mainly because they offer an additional mechanical shear or stress on the cells that are growing inside the reactor, and they can actually knock apart assemblies that have already taken place... The second thing about it is that it does interrupt some of the flow streaming that's necessary to feed the cells.' The engineers on the ground were trying to devise ways to help astronaut Thomas mitigate the problem. 'There are several strategies that you can undertake in order to shrink or eliminate the bubbles,' Dr Pellis stated. 'The first is a mechanical approach. We changed the rotational configuration of this reactor in such a fashion that it also acts like a centrifuge instead of a bioreactor, so those things of greater mass will move to

Figure 4.17. Dr Andy Thomas on Mir, and the scientists and engineers on the ground, were frustrated by the presence of a large bubble that formed in the bioreactor during the co-culture experiment. In this onboard photograph taken by Dr Thomas, the aggregate of cells can just be seen at the bottom of the round bubble that fills much of the cylinder. Analysis of the tissue upon return to Earth indicated that both cancer and cells that would form blood vessels were grown in the bioreactor on Mir.

the outside, and the gaseous bubbles that are less [massive] will move to the inside, proximal to that inner cylinder, which is actually a filter... The attempt would be then to increase the flow rate [out of the inner cylinder into the culture] and force [the bubbles] out'. That strategy did not produce the desired result.

The second approach is to 'metabolically eliminate them, and the way that we do that is that we slow the rate at which we feed the cells', Dr Pellis continued. 'When you feed the cells you feed them oxygen along with nutrient. The bubbles are made up of oxygen and nitrogen and carbon dioxide, primarily. What you do is force the cells to eat the oxygen out of the bubbles, and the bubble decreases. As it decreases its amount of oxygen, its nitrogen concentration goes up, and by what's called 'partial pressure difference', the bubbles will actually redesolve back into the culture medium, and slowly you can erode the bubble away.' But, he added, 'the extent to which we can be successful doing this is the unknown.' It eventually became possible to reduce the size of the bubbles, but they plagued Andy Thomas' co-culture experiment to the end of his stay on Mir.

One of the difficulties in coping with the bubble problem in the bioreactor on Mir, Goodwin reports, was that due to communication limitations he had few opportunities to talk to Andy Thomas. It took a few days from the onset of the problem before the scientists on the ground were informed, and they couldn't 'walk him through' the difficulties. This will change when experiments are carried out on the International Space Station, Goodwin said, because the Principal Investigator 'will be involved at every level, including on orbit.'

Mir, he concluded, was a 'positive learning experience'. On Andy Thomas' flight they determined the rates of fluid flow through the system, and have since changed the size of the tubes that are used. These and other hardware problems could be seen only in a long-duration flight. When the bioreactor had flown on the Space Shuttle STS-85 mission in 1997, there were no problems, so they would not have seen them from Space Shuttle flights.

Dr Pellis concurs. In an interview conducted on 9 August 1999, he said that the results they were able to obtain in the co-culture experiment onboard Mir were 'very preliminary'. Quite often 'you have to do experiments four and five, and sometimes even 20 times. In space you don't get that many opportunities. This was a first opportunity, albeit limited. Nonetheless, we already know that we did good assembly of tissue in space, and that two different kinds of cell can coexist in a culture system without having one prevail over the other. They can actually function in a cooperative rather than competitive fashion. That we do know for certain... There are indicators in the photographic data that, indeed, there may be some unique ordering of the cells that takes place in space, that might favour some early stages of angeogensis, or the formation of micro blood vessels. We had modelled it in the bioreactor system [on the ground] prior to that. We had an indication that we may get a favourable response in space. We continue now with the bioreactor level of this research on the ground in anticipation of acquiring another flight and eventually being able to test the hypothesis for real again.'

Goodwin mentioned that they hope to re-fly the experiment on the ISS around the time of Space Shuttle STS-106, some time after January 2001. In the meantime, in this line of research, as in all of the others, simulated microgravity research will continue on the ground.

Dr Pellis explained the relationship between the ground- and space-based bioreactor research. Microgravity 'can be a tool for us to understand the basic celluar mechanisms involved' in cell growth and differentiation. 'Once we understand the fundamentals of this, we're left with a choice: is space the only place we can do this, or can we now, knowing this mechanism, do this on the ground, and do it efficiently? This is what I call the smart use of space. The smart use of the space experience is for discovery, and for elucidating fundamental mechanisms. The question after that is, is it unique to space, or can we do this on the ground now?'

Dr Pellis remarked that he has appeared with noted heart surgeon and innovator Dr Michael DeBakey, who has been very supportive, in a number of different settings. 'We've been paired up on Congressional panels with him, and I've spent a lot of time talking with him, both formally and informally, about this. He knows that we actually have a good scientific basis for where we're doing. Plus the interesting thing about the bioreactor is, it has two modalities for operation – one that is ground-based and one that is designed strictly for space... We have, in the bioreactor on the ground, an excellent model system for microgravity. It's not perfect. No model is perfect, but it is a very good model for ripening hypotheses, and [then] putting the most robust science into space. We are less frequently putting experiments into space with a question mark, saying 'what happens to it?' More often than not, we have a declarative hypothesis; that is, making a statement based on the model, and we're going to test the model's validity in space. That's a much more productive way to do science... Dr DeDebakey told me, 'you have a very unique opportunity that a lot of other space science doesn't have.' For materials science research, you get [to use] sounding rockets and drop towers and some parabolic flights [for only a few minutes of simulated microgravity]. But we do a sustained modelling for cells that can last hours, days, weeks, months. We've had tissues growing on the ground for six months in the bioreactor continuously. That is practically unheard of as a model for profiling experiments on the ground, and translating that setting precisely as it is [into space], except that it's in microgravity now. We've been the beneficiary of that roll of the dice; we're just very lucky that this area of science can have that kind of model.'

David Wolf expressed the same appreciation for the relationship between the ground and space research: 'Space gives us certain key pieces of information', he said in an interview, 'which, ideally, we'll be able to transfer to the ground, once we know how to do it in space... Space is like cheating. We can take steps ahead in the three-dimensional organisation [of tissue] before we know how to do it on the ground, and get the information. We know it can be done on the ground, because a mother's womb does it every day. But we don't know all the secrets yet, and by doing it in space we can leapfrog many of these details, and then backfill those details [on the ground].'

4.9 LESSONS FOR THE FUTURE

After Andy Thomas' experience on Mir, Dr Pellis, and the entire team working on the bioreactor, now have a sensuous appreciation of what it will take to run long-duration bioreactor experiments in space. One lesson learned, Dr Pellis reports, is that 'in real time, if you have a problem, the system of operation aboard the space station and the communication links between us and the actual operators have to be consistent with performance of timely diagnosis. That's what we learned from Mir. You cannot wait four or five days with a bubble rolling around in the device to try to figure out what's going on. Categorically, we now know there has be a real-time response that will allow us to manage a problem if it occurs... The best way to treat anything is to prevent it so you don't have to do a treatment protocol. It's eminently easier to prevent cancer than it is to treat it. And it's no different with these kind of problems. We've done a number of redesigns based on that experience, and that was what the Mir programme was for. Some of the space naysayers, and some of the people that doubt the real value of the space programme, I think, neglect to know that this was a risk mitigation scenario. But the big gain for us was in the design of our experiments – our equipment and our procedures for our operation on space station... We had an opportunity to find out what it's like to operate this long term. Believe me, life is a lot different in the world of machinery when it only has to run for 16 days versus running 135 or 140 days. As crafty as we are in the design of a lot of our stuff, there's no guarantee until you get to microgravity and actually do this. It never ceases to surprise me. It always finds a way to get us.'

This is especially true with a bioreactor, he explained. 'One of the reasons is that our particular kinds of payload are complex; they've got a lot of different kinds of things in them, and we've got moving fluid against gas-permeable interfaces. It's a machine wearing a big 'Kick Me' sign. We've now been kicked a couple of times, and we now know what some of our preventive measures are going to be... In fact, considering the components that we've added and the things we've taken out, and the resdesign, we've even developed a new bioreactor. If you do have the untoward experience of accumulating a bubble in the reactor, the bubble is extractable now. There's an exponent of clever that comes out of these folks who work in the space programme all the time, and this was another one. It's a very clever, very simple design that allows you to do this. You can migrate this bubble to exactly where you want it, that is an open lure port, and suck it out. It doesn't require that you open up the instrument, or that you put the experiment at any risk. You don't compromise the microgravity environment... By virtue of having had the experience on Mir, we're able now to have a better chance in the station of hitting the street running with regard to the conduct of research. The Mir experience was supposed to be 'going to school'. And the last increment [with Andy Thomas] was like graduation day.'

To prepare for the next series of experiment, Dr Pellis said, 'We've tested some of this new equipment. We're doping our first space prototype configuration right now. We just got back from the machine shop the very first versions that are in the space configuration.' The team at JSC hopes that there will be more opportunities for

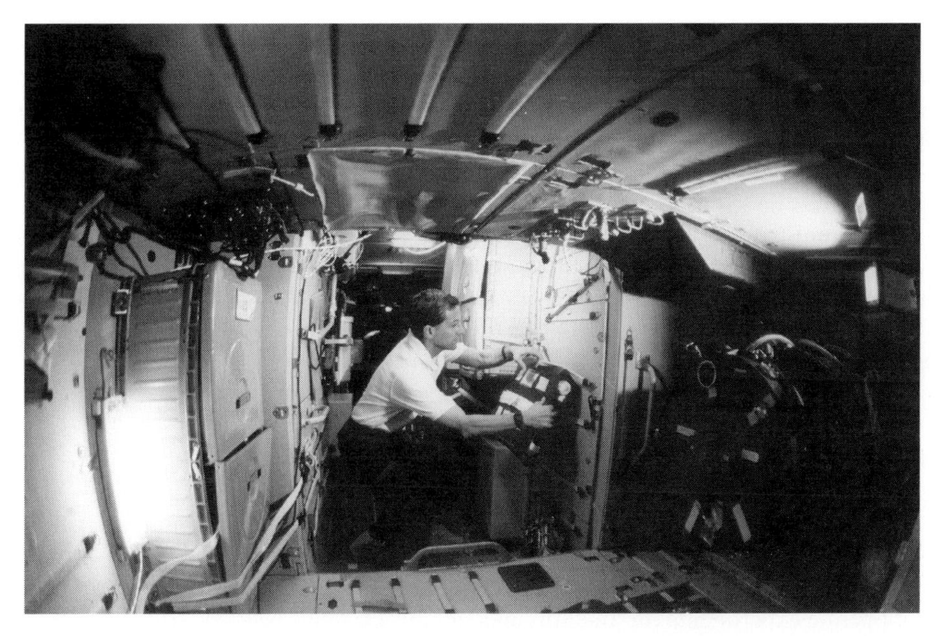

Figure 4.18. Dr David Wolf stresses that the long-duration studies in the bioreactor that will be possible on the International Space Station, and the care that a crew-member can give to the experiments will be critical to advance the research in this promising technology. In this photograph in the Priroda module on Mir, taken during the STS-79 joint docking mission, Shuttle astronaut Jay Apt prepares the bioreactor for its long-term experiments.

science on Space Shuttle missions, because they are not sure how much they will be able to do on the station during the current assembly phase.

Astronaut Wolf points to an additional number of 'lessons learned' from the shortcomings of carrying out research on Mir. 'Most of us [astronauts] came from working in laboratories, and once you're an astronaut it's a different environment. I did feel that it was a real pleasure to get back in the lab and do my work. But it was hard to do this at the same time as working with the Mir systems. I had a lot of responsibility for helping with the repair work', he said in an interview, 'and this knowledge of how to conduct the research, concurrent with the mission operations, is critical... How do we schedule people and let them work?'

He provided a specific example. 'There was a time a month and a half into the mission when I knew that to ideally image what I was seeing in the culture, I needed to set up another whole instrument from another experiment – a video microscope – but that would have taken a few days, just to get it out and set it up, and there wasn't room at the time. I would have liked to call down and say, 'Look, we need to change course here, completely change course, and take a few days out to document what we've got here.' But we couldn't do that, because we had other experiments coming on line, and we had a schedule to keep.'

After he returned from his mission to Mir, David Wolf told the NASA managers that 'if we're going to work as we do on Earth, we need to have the flexibility. Very few discoveries come on schedule. We need to let discoveries happen. We need to have a flexible schedule, and optimal use of the human that is up there making judgements and observations.'

This will not be easy, he realises. Now, 'if we [the astronauts] don't do it all, and we just don't go through the motions of doing it all, we get criticised as a failure, or for not completing the mission. On the other hand, if I make the judgement that we need to proceed with *this* work at the expense of others, you can imagine the uproar there would be from the others! This is a challenge for us – how to get this kind of flexibility into the programme and also [take into account that] people have done many years of preparation, and we need to do their experiments also.'

4.10 THE NEED FOR LONG-DURATION STUDIES

The life sciences divisions at JSC are now negotiating with International Space Station operations managers to 'find the good fits for the kinds of experiments that are in the queue versus the kinds of crew time that can be available. We probably will not operate experiments for an entire increment,' Dr Pellis believes. 'If the increment is 90 days, probably in the early days we are going to conduct four or five sequential short-term, five-day, experiments' with the bioreactor.

This is not a loss, he explained, but a gain. 'We actually come out ahead because when we go back to the old orbital format in the Shuttle, you could do one experiment. Then you had to wait 14, 16, 18 months, two years, to do a repeat. From the time of inception, when the grant was written, to the time that you get your confirmatory data from the space experiment that can be published, can be as long as six, seven, eight years. That's ridiculous. We plan to be able to do within a 90-day period [aboard the ISS], four or five confirmatory experiments, and bring back a fully analysable package of data that can then go to publication... In order to grow tissue in space, there are a lot of fundamental questions that we have to ask that require short-period experiments. We probably are going to queue those in first – the short-duration ones – and get those answers, and use that knowledge as a platform for going ahead and launching engineering of tissue in space. Taking 90 days to grow one tissue is another part of the experiment opportunities that we have.' In order to make that possible, in terms of how much crew time will be available, some of the hardware that will be used by the bioreactor is being automated. 'We have some automation to our instrumentation here already. There's another set of instruments that we will have access to, called the cell culture units, being developed in a partnership between Payload Systems Inc and MIT in Cambridge, being manufactured for Ames Research Center. This is a fully automated carousel-type device which is conducive to automatic experiments, but it's not conducive to being recharged on orbit. It has to come back.'

The suite of equipment at JSC 'is designed to remain in residence on the station for a year or more. We migrate in cell and tissue specimens and migrate out archived specimens that are either frozen or fixed for analysis, in the early station days. In

Figure 4.19. While there will be opportunities for scientists to conduct unique experiments in bioreactor systems in the microgravity of space, many more will be able to use the rotating wall vessel bioreactor to continue to conduct ground-breaking experiments on Earth.

later station days, we're going to fly some analytical capability on board so we can do *in situ* analysis.'

Ultimately, Dr Pellis hopes, there will be astronauts on the International Space Station who are trained to operate this specific hardware. But for now, and for the next few years, astronaut selection is primarily based on the abilities required to carry out the assembly phase work. 'I think as the research demand increases (and I'm very confident that it is going to increase), the demand from the science community, once it's realised what the discovery opportunities are, will be very high. The station, by virtue of being what it is – a research platform – will have to make adjustments.'

Members of the scientific community involved in bioreactor research have expressed their view of the importance of a long-term-research facility in Earth orbit. Dr David Scharp, researching cultured-cell treatments for diabetes, said in Congressional testimony, held to discuss the importance of a space station for biology research: 'Biological systems take time for adjusting to different conditions, such as microgravity... Thus a two-week maximal time of study [on the Space Shuttle] is sufficient to only provide limited information about the microgravity influence on biological systems. More thorough and productive biological studies can only be designed when these biological systems can be studied in an ongoing microgravity environment, such as that provided by a space station.' Dr Scharp also expressed his strong belief that 'biological experimentation needs to be performed by biologists on a continuous and ongoing basis, [because] unlike physics, engineering, or similar hard science experiments, in which the outcome can be readily predicted, [the results of] biological studies are often more difficult to predict... Expecting significant biological questions to be answered by limited, infrequent, short-term studies as provided on the Shuttle, is simply biologically naive.'

Scientists who have been involved with research using NASA's bioreactor technology have even devised rather imaginative potential uses for cells cultured in space. Dr Pellis, for one, proposes that in the future, cells could be sent into space as exploratory probes. Like the traditional canary, used to test the atmosphere of a coal mine, cell cultures could be designed to respond to environmental conditions of other planetary bodies in such a way that scientists could judge whether an environment is suitable for life. 'Using these probes, we could determine if the atmosphere is supportive of cells, if there is water, or if the environment is amenable to propagation.'

Perhaps Dr Jeanne Becker best described it in her article in *Earth Space Review*: 'The applications of this technology are limitless.'

5

Space crystals fight against disease

There are more than 100,000 proteins in the human body, performing a multitude of functions vital for life. Proteins are a class of chemical compounds that carry information from the DNA (deoxyribonucleic acids) that comprise the genes in every living organism, and perform an array of essential functions throughout the cells and the body. The DNA molecules cause the production of thousands of different proteins, some of which can cause or inhibit disease.

Understanding the structure of proteins can provide insight into the specific biological roles they play in the body, but presently scientists know the structure of less than 5% of them. One effective way to tease this information from a protein is to grow it in a highly ordered form, as a crystal, but this has not been an easy task in laboratories on Earth. Proteins, unlike the inorganic crystals with which we are most familiar, such as table salt, have molecular masses that range from about 890 to 2,200 times that of relatively simple organic compounds like sugar, which is easily crystallised. Dr Roger Kroes, of the Marshall Space Flight Center, describes protein crystals as made up of 'much more complex molecules. They wind themselves up a little bit like a ball of twine.'

Dr Alexander McPherson, of the University of California at Irvine, explains that 'every crystal is a collection of molecules that organise themselves so that every molecule has identically the same orientation in space and identically the same chemical environment throughout the crystal.' They also all have the same relationship to their neighbours, and form 'an ordered array of molecules.' Highly ordered crystal structures, obtained using the sophisticated diagnostic technique of X-ray diffraction, can help scientists develop methods to understand the function of proteins in the human body. 'The ultimate goal is to know the structure of the protein', Dr Kroes states, 'so you can design things to interact chemically with the protein; such things as medication. When you know the structure you can know what will interact with them – a key and lock situation.'

But large, macromolecular crystals, such as those made of proteins, are mechanically unstable, and are sensitive to temperature and other changes. Dr Dan Carter, head of New Century Pharmaceuticals, explains that protein crystals are 'usually comprised of between 30% and 80% of disordered solvent within this matrix of protein, so they're very fragile, and are held together by very few intermolecular contacts. Unlike small molecules, which have many atomic interactions which hold the crystal together, protein is a like a very large ship, each one being anchored with a few little anchors so it is kind of rocking in place. There is a lot of thermal motion and disorder which prevents the very high-resolution studies of the structure.'

Growing protein crystals requires coaxing protein molecules out of a supersaturated solution to form a coherent structure around a point of nucleation. A supersaturated solution is an extremely concentrated solution in which there is 200–1,000% more protein than can normally be dissolved in a solvent, such as water. A protein solution can be driven to supersaturation by various methods, including changing the temperature, or adding a salt precipitant, which sequesters water away from the protein, causing the solution to become supersaturated.

Dr McPherson explains that 'proteins are always grown in solution, because if they become dehydrated they collapse. They can be thought of as ordered gels, which are approximately 50% solvent. Because of the extensive channels that exist in the molecule which are filled with water, they can admit other compounds that interact with the protein or nuclei acid, such as drugs, or inhibitors,' which accounts for their promise as an intervention point to fight disease.

In a supersaturated solution the crystal will tend to re-establish chemical equilibrium if it is in contact with a less concentrated solution. 'The key to crystal growth', Dr McPherson explains, 'is that the change in saturation is gradual. If not, the process can induce precipitation, where an amorphous solid forms instead of a crystal. The standard approach is to equilibrate a highly concentrated solution with a precipitating agent. These include salts, ammonium sulphate and organic solvents. All precipitants encourage the interaction of macromolecules in solution. Salt, for example, competes with the protein solution for water, leading the molecules to interact with each other rather than with the water in the solution'.'

Up to the present time, Dr McPherson points out, 'crystallisation by any of the current methods is unpredictable. Each crystal is unique. The nucleation of crystals is still a phenomenon governed by probability rather than predetermination.'

Space has opened up a wholly new capability for growing protein crystals, which promises to allow scientists to both better understand the fundamental process of crystallisation, and to develop the ability to enhance, alter, or interfere with a particular protein's activity in the body. Freed from the limiting effects of Earth's gravity, scientists have produced crystals of proteins in space that are larger, more ordered, more homogeneous, and therefore more amenable to detailed analysis, than those grown on Earth. By observing crystallisation under these unique conditions, they will be able to discover the 'why' and 'how' of this delicate process, which has the potential to open new pathways for treating disease.

5.1 THE LEGACY OF SKYLAB

Dr Robert Snyder, now with New Century Pharmaceuticals in Huntsville, Alabama, was involved in the inorganic crystal growth experiments onboard Skylab, in the

Figure 5.1. Microgravity has provided scientists with the ability to grow larger, more perfect crystals of proteins, vital to the health of everyone. The study of space crystals – such as this crystal of insulin which was grown during a Space Shuttle mission in 1991 – has helped researchers discover the structure of these complex molecules and their functions in the body, and is leading to the development of structure-based 'designer' drugs and other tools to fight disease.

1970s, at Marshall Space Flight Center. He reported in an interview that when interest developed in protein crystal growth, at the time that the Space Shuttle could provide a platform to growing crystals in microgravity, the group of Skylab researchers from MIT, led by Gus Witt and Harry Gatos, and those who had been carrying out Skylab, Apollo–Soyuz and sounding rocket inorganic crystal experiments, came to the meetings held by NASA to discuss this new application of microgravity. It was a 'small community doing crystal growth experiments' who were familiar with each other's research. Many researchers were interested in a variety of crystal growth experiments, leading to much cross-fertilisation between the non-organic and protein crystal research. The work of Franz Rosenberger at the University of Alabama in Huntsville, for example, was 'extremely important', as he has carried out both inorganic and organic crystal growth research.

'There was a major advantage to protein crystal growth from the work with inorganic crystals', such a semiconductor materials, that was undertaken on Skylab, Dr Snyder reports. Those carrying out inorganic crystal experiments on Skylab learned that 'you must keep the temperature fairly constant when the crystals are growing'. It was necessary to 'keep the furnace at a steady temperature'. This would be a lesson for those interested in protein crystal growth, and also led to the development of techniques for temperature control.

Figure 5.2. Skylab provided the first opportunity for scientists to test their hypothesis that removing turbulence and other effects of gravity on crystal growth would improve the quality of these delicate structures. Here, at the White House on 13 December 1974, President Gerald Ford holds an encased crystal of indium antimonide that was grown on Skylab on 6 January 1974. Dr James Fletcher, NASA Administrator, explains to the President as Dr Harold Johnson, of the Massachusetts Institute of Technology, looks on.

Dr Snyder states that protein crystallisation is an area of research that is interdisciplinary, so to move it forward, early on in the programme, they 'got everybody together – crystallographers, fluid dynamicists, and so on.'

Twenty-five years after Skylab, there is now a flourishing community of researchers, from different fields of science, working to unravel the secrets of the crystal. According to the Summer 1999 issue of NASA's *Microgravity News*, 185 different proteins, RNAs, DNA and viruses have flown in space, and more than 80 investigators are involved.

5.2 THE FIRST SPACE PROTEIN CRYSTALS

The maiden flight of NASA's Space Transportation System, the Space Shuttle, took place on 12 April 1981. While the first few flights were primarily engineering missions to test all the new systems on the world's first reusable spacecraft, missions were already being planned for using the Shuttle orbiter as a temporary scientific laboratory.

The ninth Space Shuttle mission carried the European Space Agency's contribution to this new manned space programme – a pressurised laboratory. Spacelab housed dozens of scientific and technology experiments, and during a ten-day mission in November and December 1983, astronauts tried out the new laboratory.

One of the 70 experiments, designed by Walter Littke, Professor of Chemistry at Freiburg University in the Federal Republic of Germany, was called 'Crystal Growth of Proteins'. It was the first space experiment of its kind. During the 1970s, while astronauts aboard Skylab were growing inorganic crystals, with dramatic results, Professor Littke was also growing crystals in his laboratory at the university. He noted that his earthly protein crystals would break, and were fragile and small. He postulated that this was due to the effects of gravity. It is not surprising that a German scientist would be working in this field. *Microgravity News* reports in its Summer 1999 issue that the 'first known published observation of the crystallisation of a protein was made by F.L. Huenefeld at Leipzig University in 1840.'

A decade before Spacelab 1, during the Skylab programme, scientists postulated that the beneficial effect of removing gravity from the process of crystal formation would apply not only to inorganic materials, as had been demonstrated, but also to various proteins produced by living systems. Scientists hypothesised that there are two principal ways in which gravity interferes with the crystallisation of macromolecules such as proteins.

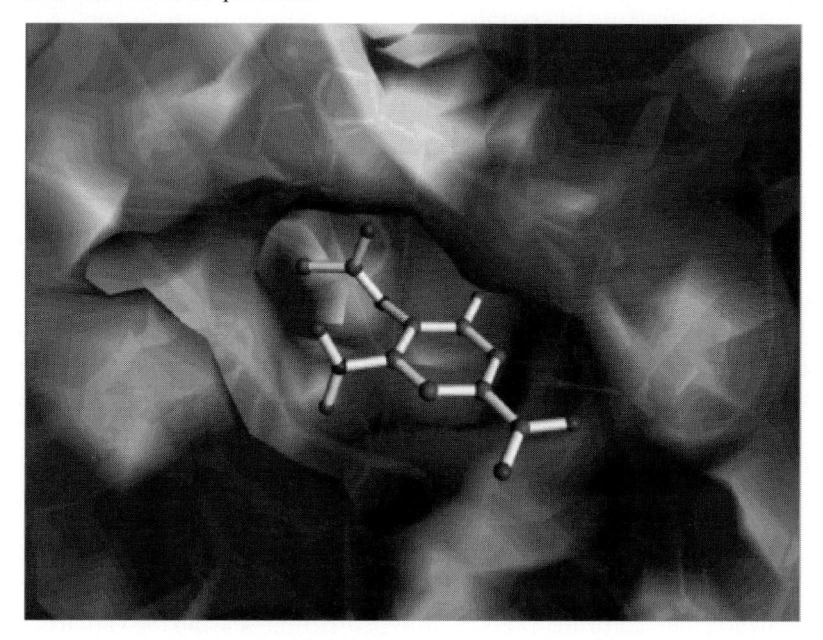

Figure 5.3. The purpose of trying to grow larger, more perfect protein crystals in space is to determine the geometric arrangement of the atoms and molecules that make up the crystal. This artistic rendering of a virus crystal grown in space demonstrates the challenge of 'teasing' the structure from the cloud-like molecule of a protein.

Under ordinary circumstances, when a crystal is forming from a solution, material near the crystal is less dense than the surrounding fluid, as protein molecules collect to form the crystal. This now less dense material rises, creating gravity-induced convection currents in the solution and near the forming crystal. The convection currents prevent the formation of stable depletion zones around growing crystals. Disorder is believed to be the result.

A second gravity-induced effect that interferes with crystal growth on Earth is caused by the weight of the crystal. As it is adds molecules and becomes heavier, it can drop to the bottom of the container, through sedimentation. If the crystal makes contact with the wall of the container, the shape can be distorted.

Before trying his experiment in microgravity crystal growth in Spacelab on the Space Shuttle, and in order to test his hypothesis that a diminution of gravity would improve the ordering of protein crystals, Professor Littke conducted preliminary short-duration (6-minute) experiments with sounding rockets in the European TEXUS programme. Images he received of the crystal growth process during those few minutes of microgravity convinced him that the process was not the same under Earth and space conditions, he and co-investigator Christina John reported in 1984.

At a three-day Symposium to discuss the results from the Spacelab 1 experiments, held in March 1984 at Marshall Space Flight Center, Dr Karl Knott, of the European Space Agency, reported that Dr Littke's apparatus had produced protein crystals 1,000 times larger than samples grown on Earth. This will allow better study of the molecular structure of crystals, using X-ray diffraction techniques. Knott was optimistic that such space crystals would have potential uses in molecular biology research.

In the 13 July 1984 issue of *Science*, Dr Littke and Christina John reported on the results of their protein crystal growth experiment on Spacelab. Using a liquid–liquid diffusion system, the experiment produced space crystals that were 27 and 1,000 times larger than those exposed to terrestrial gravity. The protein that showed the greatest gain was lysozyme, a protein found in egg whites. Lysozyme is often used as a model to evaluate crystal growth processes, because its structure is well known, having been discovered 17 years before Dr Littke's experiment flew in space. Their analysis showed that the effects of convection and turbulance were observed in the crystals not in space, but 'only when the payload began to re-enter the gravitational sphere of the Earth'.

Dr Littke's experiment quickly created interest in space-based protein crystal growth in the scientific community in the United States.

One of those most excited about the possibility of this new tool for science was Dr Charles Bugg, Professor of Biochemistry at the University of Alabama, Birmingham, Medical School, and Associate Director of its Comprehensive Cancer Center. Dr Bugg, and Drs F.L. Suddath and Larry DeLucus, who were faculty members of UAB, designed the first protein crystal growth experiment using the vapour diffusion method to fly in space, which it did in 1985. They launched almost 50 vials of ten different proteins on Space Shuttle mission STS 51-D. The most successful crystal came from a lysozyme solution, which was ten times the size of any previous crystal.

By the mid-1980s, NASA scientists were encouraged enough by the preliminary

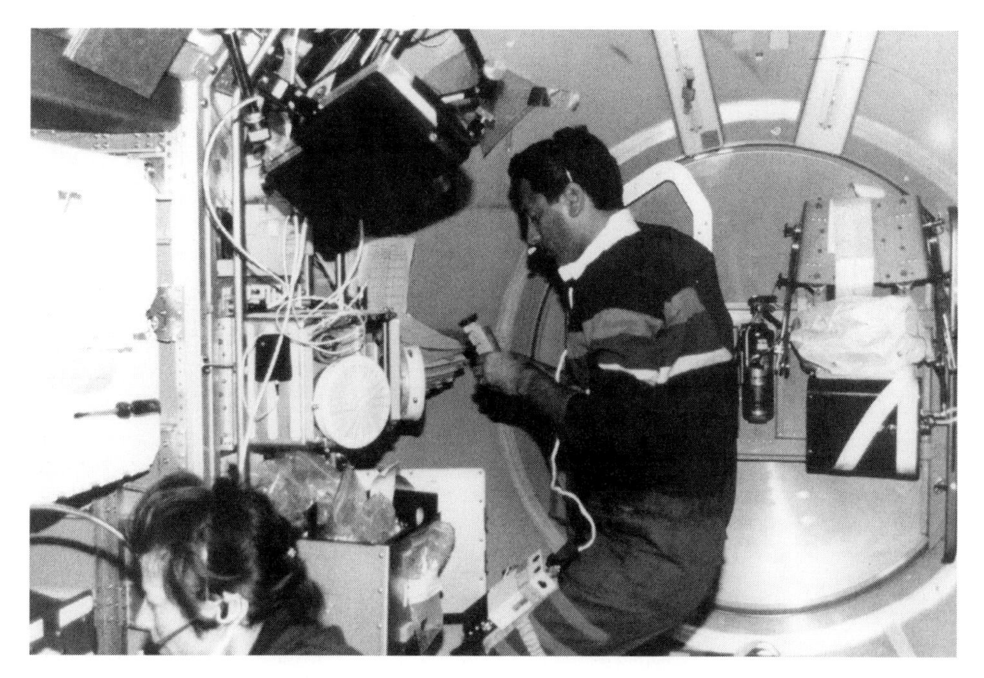

Figure 5.4. Dr Larry DeLucas, the current head of the Center for Macromolecular Crystallography at the University of Alabama at Birmingham, was fortunate to be a payload specialist on the STS-50 mission in 1992, and was able test the protein crystal growth equipment in space. Astronaut Bonnie Dunbar (lower left) communicates the status of the experiment to the ground.

results of protein crystal growth experiments in space, that in 1985 they established the Center for Macromolecular Crystallography at the University of Alabama at Birmingham. The purpose was to 'form a bridge' between the space agency and private industry, in order to stimulate biotechnological research in space. The Center grew out of a joint venture between NASA and industry to develop protein crystal experiment hardware for Space Shuttle flights.

Between 1985 and 1986, experiments on four Space Shuttle missions were conducted by the new Center, to develop and verify a vapour diffusion technique for growing protein crystals in space. By 1988 the scientists were ready for a full-fledged experiment, and on Space Shuttle mission STS-26 in September 1988 the vapour diffusion method was used for a series of protein crystal growth experiments. Investigators from the Center for Macromolecular Crystallography, NASA's Marshall Space Flight Center, E.I. du Pont de Nemours & Company, Upjohn Company, Merck Sharp & Dohme Research Laboratories, Schering–Plough Research Institute, the University of California at Riverside and other academic and commercial institutions, participated in the experiments.

In the vapour diffusion method, a small volume of protein solution is allowed to equilibrate with a precipitant solution through the vapour phase. As it moves toward equilibrium, protein molecules form a crystal. In the 1988 Space Shuttle experiment,

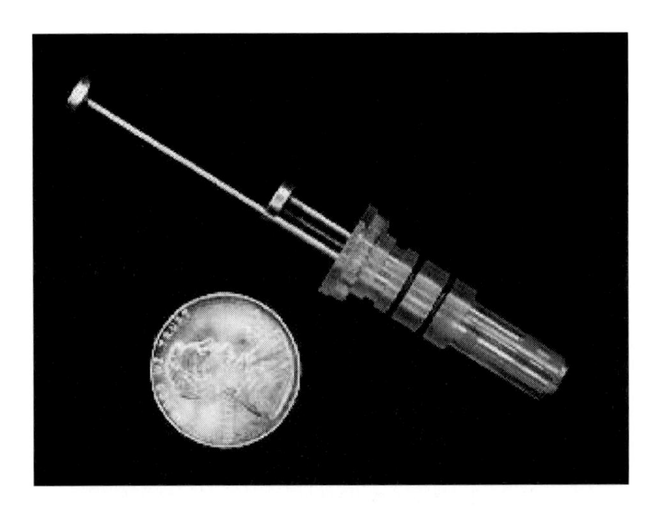

Figure 5.5. The second-generation Vapour Diffusion Apparatus uses a triple barrel syringe to grow protein crystals in microgravity. Two barrels of the syringe hold the protein and precipitant solutions until they are mixed at the top of the syringe. The third barrel pulls the drop in and out to mix the solutions. The coin is included to provide a scale.

a double-barreled syringe, contained in a chamber, was loaded with protein solution and with precipitating solutions, in adjacent barrels. Once in orbit, crystal growth was activated by extruding the two solutions to form a small drop on the tip of the syringe, where they mix through repeated withdrawing and extruding of the solutions. The droplet is suspended at the tip of the syringe, and the chamber surrounding it contains an absorbent material that is saturated with a more highly concentrated solution of the precipitating agent. To reach chemical equilibrium, water leaves the suspended protein droplet and travels as a vapour to the reservoir of precipitant. Almost half of the water in the protein moves to the reservoir as the crystal begins to form.

The growing crystals were photographed at 24-hour intervals during the experiment. After three days the solutions and suspended crystals were withdrawn into the syringes, and sealed for return to Earth. Identical equipment was used for control experiments on the ground, which were actually started seven days *after* the Space Shuttle landed, so that the conditions experienced in space could be reproduced in the control, except for the absence in space of the affects of gravity.

The technique used to analyse the results of the experiment is called X-ray diffraction, in which the crystal is bombarded with X-rays, which excite the electrons on the outer structure of each atom to produce a pattern of scattered X-rays. This has to be repeated many times from different angles in order to produce a complete pattern. From these data, researchers build an electron density map to determine the atomic structure of the protein, and can compose a computer-generated model of the structure. With such information it is possible to locate receptor sites and active

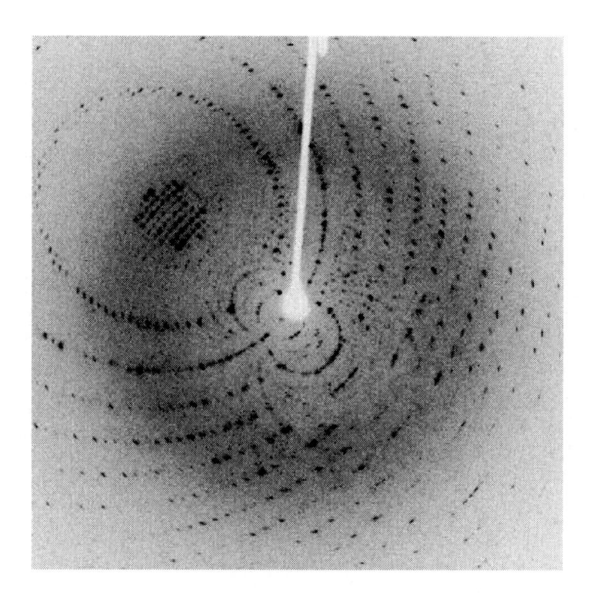

Figure 5.6. X-rays are commonly used to illuminate crystals so as to be able to create crystallograms from which the structure of the protein can be deduced. These dots and shadows from an X-ray crystallogram of lysozyme reveal the inner structure of the crystal. Newer techniques, such as neutron diffraction, are adding to the information that can be gleaned from space crystals.

areas that control the functions of the protein. One drawback of the X-ray diffraction method is that the multiple exposures generally damage the crystal.

X-ray diffraction photographs were produced for a qualitative evaluation of the results of five crystallisation experiments involving 11 proteins onboard the Shuttle. Two crystals that formed were larger than the best that had been produced in any ground experiments, and one of them was as much as 50% larger. Of the other eight proteins, six did not produce crystals large enough for diffraction analaysis, which the experimenters attributed to non-optimal crystal growth conditions such as the lack of sufficient growth time (crystals grow more slowly in microgravity); one was only partially successful, and one suffered difficulties during processing.

Although not a total success (as is the case with most experiments) the scientific investigators stated that 'when the results of these experiments are considered together, the data indicate that protein crystals grown under microgravity conditions diffract to higher resolution than the best crystals obtained under similar conditions on Earth'. They concluded that 'the ability to grow protein crystals with increased molecular order should enhance crystallographic solutions by shortening the time required to determine the structure and by increasing the accuracy of molecular details.'

Because this application of microgravity processing seemed so promising, even from very limited flight experience, a community of researchers developed, from the government, academia, and industry, to pursue different lines of approach to growing protein crystals in space. The goal was to understand the fundamental process of crystal formation, to unveil the structure and organisation of the crystals, and to use this new knowledge to gain new insights into the role of proteins in disease, treatments and cures.

In 1991 NASA started a research programme, led by its Marshall Space Flight Center, to fund a variety of approaches to this research, with the goal of establishing the fundamental physics of crystal growth.

5.3 MIR OPENS A NEW WINDOW

Dr Dan Carter, the founder of New Century Pharmaceuticals, joined NASA in 1985, when protein crystal growth was just entering the Space Age. In an interview he explained: 'Around 1991 [a group of us as at Marshall] submitted a proposal for a different idea on how to do protein crystal growth. It was a new concept with the possibility of getting vapour diffusion, but with much greater numbers of experiments. We started designing a separate experiment in 1991, 1992, and actually flew it in 1994. This was a vapour experiment on the Space Shuttle.' (In 1996 Dr Carter was appointed as the chief scientist for biophysics for NASA at MSFC, but left shortly afterwards to found New Century Pharmaceuticals in January 1997.)

In 1995 there came a new opportunity to grow protein crystals for weeks or months in space, rather than days, for the first time. Dr Carter explained that his experiment onboard the Russian station used the liquid–liquid diffusion method –

Figure 5.7. The Diffusion Controlled Apparatus for Microgravity (DCAM) was designed by Dr Dan Carter and his team, specifically for flight on Mir. For Mir operation, the device required virtually no attention from the crew. The protein sample is placed in a button (lower left), which is one end plug of the cylinder, and is covered by a semipermeable membrane, which allows precise control over the rate at which the protein crystal forms out of solution. Two reservoirs of precipitant solution are separated by a gel plug, seen inside the assembled cell on the right, and by varying the position of the plug the scientists can control the rate of crystal formation.

rarely used in the gravity environment of Earth, because the direct mixing of two liquids of different concentrations amplifies the deliterious effect of gravity on the delicate process of growing crystals.

Describing his experiments on Mir, Dr Carter reports: 'Those are some of the most interesting experiments because, for the first time, we began to look at deliberate long-term experiments. Being able to do these experiments, and designing a piece of hardware, simple as it was, to conduct a long-term experiment in a controlled way, with limited human interaction, if any, has turned out to be a very nice piece of hardware. It was quickly designed to come up with a way to take advantage of these unique opportunities during the Mir increment series [with US astronauts]... From that, we've learned how to reproducibly grow almost any protein crystal to extremely large size'.

Taking advantage of these large-sized crystals grown on Mir, Carter and his group are developing the use of neutron diffraction techniques to supplement the analysis possible using X-ray diffraction, to peer into the structure of protein crystals. 'Neutrons interact with the nucleus of the atoms instead of the diffuse electron clouds as in X-ray diffraction, allowing us to see things like the precise location of hydrogen atoms in water molecules, and the active sites of proteins', says Dr Carter, 'but for neutron applications, where you're bouncing neutrons off the crystals, you really have to have extremely large crystals, around a multi-millimetre to a centimetre size. Previous studies of neutron diffraction have more or less been done because a crystal of a particular protein just happens to grow large enough, by accident.'

Neutron diffraction is a relatively new diagnostic technique. 'There are about twelve neutron structures that have been determined, in the history of science, for proteins. We have shown six examples of enablement. We finished one, we have two more that are in the refinement stage, and we hope to collect a new fourth data set this fall', Dr Carter stated in the summer of 1999. 'So we're adding 30%, and hopefully eventually 50%, very quickly to this area, and it's exciting because its scattering factors are different... The crystals don't decay in neutron beams, and you can see hydrogen atoms in ways that you can't see them in X-ray work. In some of our future experiments we'll be looking at model systems for drug design to see if this is really going to provide unique information over the X-ray structures. It's still research. I'm pretty excited about it.'

In addition to the production of large improved crystals onboard Mir, Dr Carter reports that the scientists were able to observe some of the fundamental process of crystal growth, and add to the basic theoretical understanding of this complex process. 'An area that we gained a lot of insight in on Mir, is what the role is of microgravity in improving the quality, or perfection, of protein crystals. That was one of the things, aside from the neutron diffraction area, that I feel is an important consequence of our Mir experiments.'

Dr Carter, his team, and an international group of scientists, have been looking at what is called 'impurity partitioning' in macromolecular crystals grown in space. 'Even if our crystals were damaged, which happened on a variety of flights [on Mir] – for example, when they had the Spektra collision, and our crystals were exposed to

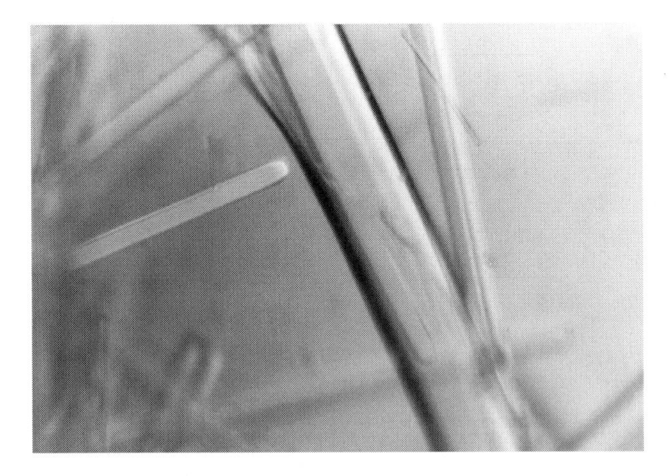

Figure 5.8. This crystal of a nucleosome core particle was grown in the DCAM apparatus onboard Mir. Because of its ease of handling, and positive results, the apparatus was flown on almost every Mir mission.

very low temperatures and they didn't have the quality they should have – as far as impurities go, since this happened after the crystals had grown, we could see the impurity profiles. We still gained a lot of information from those crystals,' he reports. 'We were able to see that, because of the absence of convection, certain types of impurity are eliminated, or greatly reduced in the crystal in space versus the crystal on the ground. As a consequence you get much more highly ordered crystals and the crystals can grow larger, and this is consistent with all of the observations people have been recording from protein growth in space.'

In the Mir experiments, one material under examination was hen egg white lysozyme (HEWL), which is easily crystallised and inexpensive, and was grown throughout a series of missions. Taking advantage of the fact that Dr Carter and his team had experience with two different methods of protein crystal growth, the crystals were produced by both the vapour diffusion (PCAM) process, and the Diffusion Controlled Crystallisation Apparatus for Microgravity (DCAM) dialysis method, the latter of which had been designed specifically for Mir.

The requirements for an experiment on Mir were that it needs no electrical power and virtually no crew time. This meant that the experiment was on its own. During his interview Dr Carter explained that the new DCAM technology allowed the experiment to be controlled 'precisely over a period of several weeks to several months. The hardware didn't exist to do that [before Mir], simple as it was.' Because of the way DCAM was designed, 'when you loaded the hardware, it was operating. We didn't have to have an activation performed in space, and we didn't have to have deactivation [by an astronaut]. That control process allowed us, over a long period of time, to grow these extremely large crystals,' without any demands being placed upon the Mir crew.

In an article in a 1999 issue of the *Journal of Crystal Growth*, the scientists describe their DCAM apparatus. It is a 'micro dialysis counter-diffusion apparatus,

Figure 5.9. The Protein Crystallisation Apparatus for Microgravity (PCAM), developed by Dr Carter and his team, uses the vapour diffusion method for growing protein crystals. A protein sample is placed in each small well, or vapour equilibration chamber, in the tray. The well is surrounded by a doughnut-shaped reservoir. The two solutions are separated by a rubber seal, which is retracted by an astronaut to begin crystal growth. PCAM has been flown on the Space Shuttle.

developed in response to special exploratory opportunities for long-duration protein crystal growth experiments' on Mir.

The experiment housing has two cylindrical reservoir chambers joined by a tunnel. The smaller chamber is the secondary precipitant reservoir, and the larger chamber is the primary reservoir. The protein sample is placed in a button which is attached as an end-cap to the smaller secondary precipitant reservoir, in which solution is less concentrated than that in the secondary precipitant chamber. The protein button is covered by a semipermeable membrane that allows the precipitant solution to pass into the protein solution, increasing its concentration and initiating the crystallisation process. The two reservoirs are separated by a gel plug of varying proportions.

The length and diameter of the plug is a means of passively controlling the rate of diffusion. By varying the dimensions of the gel plug, the investigator can allow the equilibrium profile, or the rate of diffusion into the protein solution for a specific protein, to be selected to induce crystallisation within a few days or a few months. The experiment is activated when the samples are loaded into the housing on Earth, but the rate of diffusion is so slow that no appreciable growth occurs until the samples reach orbit a few days after launch.

The first set of specimens returned from Mir were onboard STS-79 in September 1996. Scientists were concerned that the apparatus' longer than planned six-month stay (along with Shannon Lucid) on Mir would have possible undesirable effects. But when the first crystals came back from Mir, NASA reported that the visual observations had yielded 'dramatic results'. Large crystals of the nucleosome core particle, which regulates genetic activities in the nucleus of the cell, were produced. The largest T7 RNA crystal ever produced came back from Mir.

A wide variety of crystals was flown onboard Mir in the DCAM. The results, with all the problems on Mir, verified this new approach, and it is now also being used in ground-based applications. DCAM flew on five out of the seven US long-term increments on Mir, supporting more than 700 crystal growth samples. Both the experimenters and NASA wanted to fly the apparatus as much as possible, to take advantage of the astronaut increments on Mir, and so samples were grown in back-to-back experiments. Because it was not possible to check the results of the experiments as they were taking place on Mir, a new group of samples had to be prepared and ready to go before those on Mir could be brought back and analysed. This presented a challenge to the investigators.

Dr Carter explained in an interview that 'not getting the feedback on your experiments until the next cycle means you skipped a cycle every time. This was a very interesting process. Basically you couldn't impact the immediately adjacent experiment. You had to skip one. As you move to the International Space Station you'll be able to have more interaction.'

The hypothesis the researchers hoped to test in their Mir experiments, based on theoretical analysis, was that since there is a lack of convective eddies around a crystal growing in microgravity, the solution around the growing crystal would be more stagnant than on the ground, with less impurity material accessible to the crystal. This condition, they surmised, would lead to 'impurity partitioning' in the microgravity-grown crystal, where fewer impurities would be incorporated into the crystal. If there is less stress induced in the crystal from the impurities, then this could account for the fact that there is often a larger and more perfect formation of crystals in space. Crystals from three consecutive Shuttle–Mir flight experiments were examined. The studies of the crystals grown in the DCAM, where the slow crystal growth is passively controlled, and the solutions had been deliberately doped with seleced impurities, revealed a 'consistent and dramatic exclusion of the protein dopants for both ground and flight experiments.' The crystals grown in the vapour diffusion apparatus, the PCAM, with a higher growth rate, 'exhibited a greater degree of variability in protein impurity incorporation and quality.' Also interesting, the authors point out, is that unpurified hen egg white lysozome, used in the DCAM experiments, had a lower microheterogeneity when compared to ground-based crystals. They characterised this as a 'self-purification' process during crystal growth.

From the data gathered after crystal analysis, the scientists concluded that binding stresses are produced in a crystal when there is a replacement of regular protein molecules by another species, or impurities. 'The more ordered crystal structure and morphological features may be partially or wholly a product of less impurity trapped by the crystal when impurity trapping is essentially determined by diffusion.' Other groups of scientists are investigating other aspects of crystallisation in microgravity to understand the fundamental process at work, and how the space environment can lead to new avenues of applications.

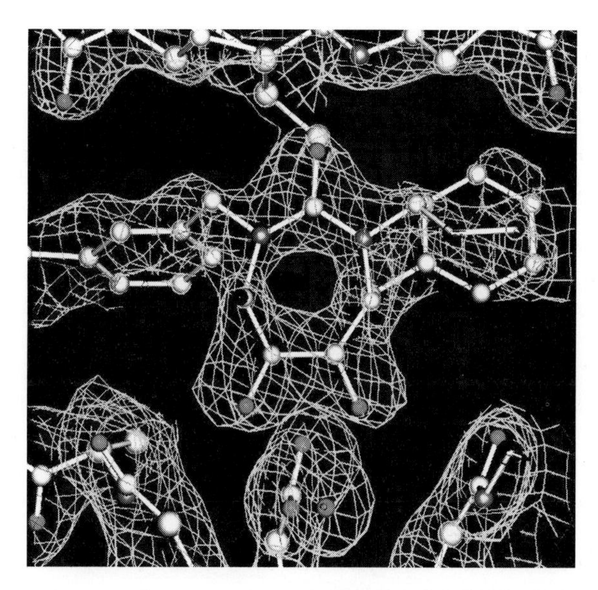

Figure 5.10. New Century Pharmaceuticals, founded by Dr Dan Carter, is working with commercial companies to help find cures for devastating diseases. HIV protease helps the AIDS virus resist drugs, and scientists are interested in finding a protein that will block or inhibit its function. Ground-based research had not allowed scientists to observe the shape of the combined proteins, to discern the effectiveness of one inhibitor that has been developed. However, this model of mutant HIV protease coupled with an inhibitor, grown in the PCAM in space, allowed scientists to determine the shape of the combined proteins.

5.4 A MULTI-USER SPACE CRYSTAL FACILITY

While gravitational convection effects made ground-based experiments using direct liquid–liquid diffusion very difficult until the development of the DCAM apparatus, vapour diffusion techniques have been in use in laboratories for several decades. Here, diffusion in the vapour phase between two liquids mitigates the effect of the direct confrontation of liquids with differing concentrations.

In response to the early-1990s NASA initiative to broaden the approaches to protein crystal growth in space, the researchers who would later form New Century Pharmaceuticals developed the vapour diffusion Protein Crystallisation Apparatus for Microgravity, or PCAM. In this device, a protein sample drop is placed in a small well surrounded by an absorbent doughnut-shaped reservoir, which forms a vapour-equilibration chamber. The two solutions are isolated by a synthetic rubber seal before launch and after crystal growth is completed. On orbit, an astronaut retracts the seal that separates the two solutions, which allows vapour diffusion to begin. Seven vapour equilibration chambers are held on a tray, and nine trays are placed in a cylinder, 81 mm in diameter and 381 mm long, which then holds 63 individual experiments.

Figure 5.11. Dr Dan Carter, founder of New Century Pharmaceuticals, in his laboratory in Huntsville, Alabama.

Over the course of 1995 and 1997, more than 3,000 individual PCAM protein crystal growth samples were flown on the Space Shuttle, with the most recent flight on STS-95 in November 1998. Typically, a manifest includes about 630 experiments shared among 10 to 14 research groups. Each PCAM flight contains 378 vapour equilibration experiments in a mid-deck locker on the Space Shuttle. Because PCAM requires a well-controlled thermal environment, Mir was not the best place for the facility. 'On Mir,' Dr Carter explains, 'there was always ambient temperature [fluctuation], meaning the experiment was always in jeopardy because we can't stand large fluctuations, which occurred on almost every Mir flight. When you're talking about a period of five or six months [temperature fluctuations are] inevitable.'

Now, post-Mir, Dr Carter is looking forward to flights on the Space Shuttle and the International Space Station. 'Our vapour diffusion hardware is a multi-user facility that we fly. We have a group of about 16 to 20 different co-investigators from labs around the world and government and industry laboratories, and most of those experiments, probably 90%, are geared toward producing high-quality crystals for direct application in structural biology.'

Now that the technology has been proven on Shuttle flights, Dr Carter said, 'we're making another step, and we have been working on this for almost two years. We are being given our own commercial programme (which we will start flying very shortly) of six to eight Shuttle flights or space station experiments over the next two years that are separate from our science base. On the commercial side, the companies can enter in, and their proprietary concerns are taken care of, and we're hoping that will open up the commercial side of the microgravity programme.'

In a 1999 paper, 'PCAM: A Multi-User Facility-Based Protein Crystallisation Apparatus for Microgravity', it is stated that development of this hardware was undertaken to adapt the commonly available vapour diffusion crystallisation hardware 'to increase sample density, and to improve flight logistics and handling.' An early hand-held version of PCAM flew on STS-62 in March 1994, as a proof-of-concept experiment to test the viability of the hardware. On that mission, astronaut Pierre Thuot photographed a growing protein crystal, and the data was downlinked to the ground during the mission. The authors summarise a selection of results from the experiments that have flown so far. In one experiment in 1995, DuPont Pharmaceuticals Company flew an HIV protease complex with a proprietary inhibitor that they have developed to treat the disease. The purpose was to improve the quality of crystals to aid in the design of new inhibitors against HIV, and to improve the potency of current inhibitors. The experiment produced the highest quality crystal of this specific inhibitor protein complex yet obtained. The crystals did not show the typical significant decay during X-ray diffraction analysis, 'which was a marked difference from the terrestrial samples.'

Another crystal grown using the PCAM device, on the STS-67 Space Shuttle mission in 1995, was of human antithrombin III. The physiological function of this material is to control blood coagulation in human plasma. It achieves this by forming inhibitory complexes with thrombin and other coagulants in a process. When there is a mutation in antithrombin, maladies such as pulmonary embolism and thrombosis can result. The crystals grown in space were superior to previous products, and revealed regions in the crystal, previously unseen. The improved understanding of the structure of this protein will aid in the design of preparations for intervening in the process of disease.

In 1997, on Space Shuttle mission STS-85, co-investigators Drs Carter and Glenn Pilkington, from Intracel, crystallised an antibody for respiratory syncytial virus. RSV is an influenza-type virus which produces serious respiratory infections in infants, and is especially dangerous to those with other health risks and complicating factors. There are approximately 120,000 hospital admissions per year in the US, with a mortality rate of 4,000. Intracel has developed a recombinant antibody which neutralises all known variants of the RSV virus. New Century Pharmaceuticals has determined the atomic structure of this antibody, and crystals grown on STS-85 were the largest to date. X-ray diffraction data from them will further improve knowledge of the structure, and development work on a vaccine. Respiratory synctial virus antibodies were flown again on 3 November 1998, on the STS-95 Space Shuttle flight with former Senator John Glenn.

In addition to producing singularly improved samples of protein crystals, and because of its ability to accommodate many samples in one flight, PCAM is also an effective method for screening large quantities of proteins for later detailed study.

The researchers envisage many applications and a contribution to science and health for this technology. As Dr Carter stated in an interview: 'In molecular biology, people are gaining a lot of insight into processes that in some cases save them a decade of work. They've been working on these problems for ten years, and didn't have this resolution' that can obtained in space.

5.5 CRYSTALS IN A THERMOS BOTTLE

After the United States and Russia signed the agreements for the Space Shuttle astronauts to live on the Mir space station, Dr Arnauld Nicogossian, head of the life and microgravity sciences office at NASA headquarters in Washington, put out a call to all the investigators in the protein crystal growth programme. Dr Alexander McPherson related in an interview that Dr Nicogossian had said: 'We need you to propose an experiment for Mir. But it can't use any electricity, it can't use power, it can't use any water, you're not going to get any crew time, and by the way, we're not going to give you any money to do it.'

'We proposed this experiment,' Dr McPherson explains, 'which used a piece of NASA hardware already in existence – the liquid nitrogen dewar [GN2 dewar], that required no power, no water, no crew time, no nothing. NASA flew it seven times on Mir because it exactly matched the requirements. Every time NASA had an empty space on one of the flights, they would call us and say 'Can you fly that Dewar?' '

The dewar is a container that can keep material inside, such as a protein sample, frozen at cryogenic temperatures, where gases can assume a liquid form: a highly sophisticated thermos bottle. In Dr McPherson's experiment the proteins are placed into small plastic tubes in which they are flash frozen, and the precipitating agent is then placed on the other end, and subsequently frozen. On the ground, the material can be kept at –180° C by replenishing the liquid nitrogen coolant as it boils off. As the liquid nitrogen bubbles off and is not replaced on orbit, the samples slowly return to room temperature. When they become liquid, the protein solution and precipitating agent slowly diffuse into each other.

Figure 5.12. Designed for flight on Mir, Dr Alexander McPherson's liquid nitrogen dewar appears to be what it is – a large thermos bottle. Inside the aluminum container, which is surrounded by liquid nitrogen to keep the material inside frozen, are three sample bundles of proteins. Up to 500 protein samples can be carried in each dewar, and crystals form as the samples thaw over time.

'It was an idea that we had, but we'd never had a chance to fly it before,' Dr McPherson explained. 'The reason that it's never flown [on the Space Shuttle] is that it takes about 12 days for [the frozen] samples to thaw out, so if you flew it on the Shuttle they would only start thawing about the time it was time to go home, and you'd get no time to crystallise. The great advantage of Mir that made it attractive for this kind of experiment was the long duration, where it was up there for three or four months, so 12 days of thaw time meant absolutely nothing. You still had three months left for the crystallisation to occur.'

Dr McPherson and his team proposed that inside the liquid-nitrogen cooled dewar could be a series of experiments to grow protein crystals using two different methods. One of these was batch crystallisation, the oldest and simplest method for growing crystals. As Dr McPherson outlined the process: 'You simply mix two things together and that produces, right there, instantaneously, a supersaturated solution. It means you make one solution and it is supersaturated from the beginning. That's not always possible. [This method is quite different from] the equilibration methods, like vapour or liquid–liquid diffusion. In those cases, you're slowly mixing, by some means, two different components, which, when they reach a certain level, produce a supersaturated solution. By controlling the rate of diffusion, you control the rate of approach to supersaturation.'

'A lot of times, if you try to do a batch experiment,' Dr McPherson stated, 'where you create a supersaturated solution right from the very start, what happens is that [instead of nucleation and formation of a crystal] you just get precipitated protein in an amorphous form. It comes immediately out of solution as a precipitate and you get no crystals out of that at all. If it doesn't happen you can use the batch method. It's nice because it's so dead simple. You just mix two solutions.'

Whether or not batch crystallisation will work in a given protein solution is difficult to predict, Dr McPherson says, because 'nucleation is a statistical effect that we don't understand. There are things happening in solution that are extremely complex, and whether you get nuclei or precipitate depends on the nature of the components and the way they interact. That varies from one protein to the next, and there are thousands and thousands of proteins. We just try it, and if it works, it works. Most of our protein crystal growth research is very, very empirical. We have some theory to guide us but basically it's a very empirical process so you just try lots and lots of things... That's why, if nothing else, the Mir experiments that we did were particularly important, because we showed that you can carry out a huge number of experiments. We flew up to 200 samples at a time, but we could have flown up probably 10,000 samples at a time in that same apparatus. We had to make the samples on the ground. We could have made 10,000 small samples which would have taken us probably six months, or we could have made 200 big samples, which took us about a week and a half. In the interest of time, we made bigger samples and filled up the container with them.'

The second method used in the dewar onboard Mir was a liquid–liquid diffusion method of crystallisation. But unlike other such methods devised for space applications, the two solutions of protein and precipitant were physically separated merely by being frozen in place. When they thaw, they simply slowly diffuse at their

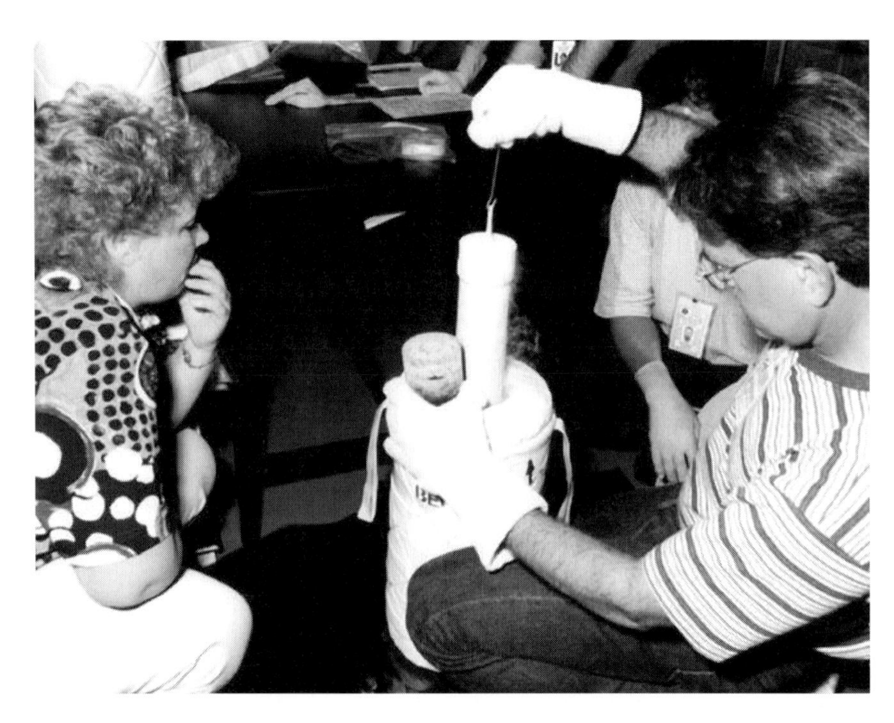

Figure 5.13. Dr Stan Koszelak, from Dr McPherson's laboratory, removes the experimental crystal growth container from the transportation dewar at the Kennedy Space Center before its launch on the Space Shuttle in 1995. On the left is Tamara Chinareva, from Mir's operator, Energia, the Russian Spacecraft Corporation.

interface. The experiments in the dewar required absolutely no human intervention. All the astronaut had to do was move it from the Space Shuttle to a secure spot on Mir.

The first dewar experiments on Mir were launched onboard the Space Shuttle in June 1995, and returned to Earth that November. Reporting on this initial proof-of-principle experiment, Dr Stan Koszelak – a colleague of Dr McPherson's at the University of California, Riverside, where many of the experiments were carried out, and now at Dr McPherson's laboratory at the university campus at Irvine – wrote in 1996 that 19 types of protein and virus samples were tried, and 17 crystallised. They observed that the small samples yielded the best results and that longer diffusion times were clearly superior. The scientists concluded that both the batch and liquid–liquid crystallisation methods had been demonstrated to produce impressive results in space.

One important result obtained during the series of experiments on Mir, Dr McPherson reports, was with beef liver catalase. 'We grew some very, very large good crystals that we collected the data on at the synchrotron facility, and we used that to extend the refinement. Beef liver catalase is a detoxifying enzyme. It's fairly important; it helps protect you.' Another experiment on Mir incorporated plant haemoglobin. 'This was interesting because it was one of the few protein samples

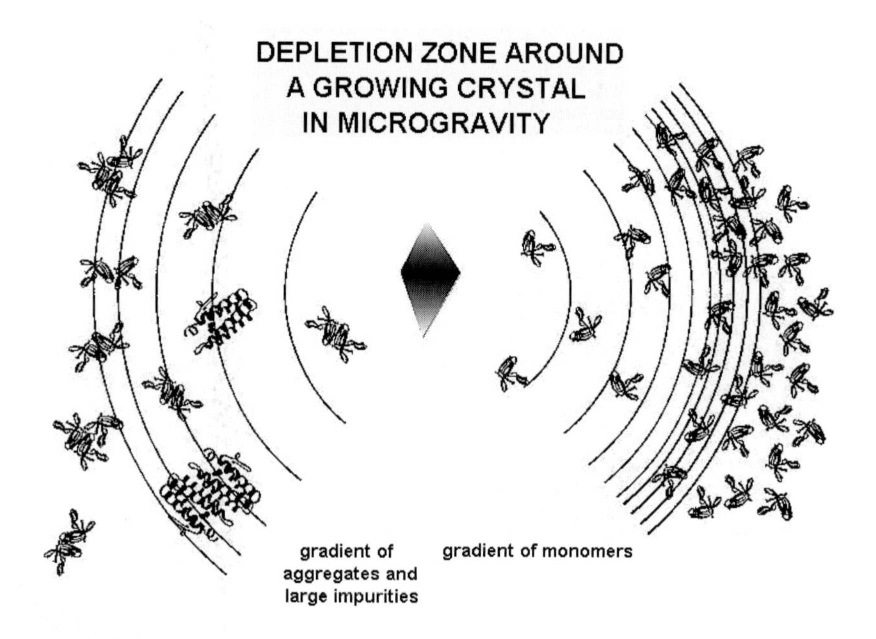

Figure 5.14. It has been observed that when a crystal is grown from a solution, a concentration gradient, or depletion zone, forms around it and extends some distance from its surfaces. The concentration of the nutrient for the growing crystal is lower near the crystal, (on the right of the diagram), diffusing more slowing and slowing the rate of growth of the crystal. Likewise, the large impurities in the solution, (seen on the left of the diagram), diffuse even more slowly than the nutrient. Dr McPherson proposes that this accounts for the reduced amount of impurities in space-grown protein crystals. (Courtesy Dr Alexander McPherson.)

that our Russian colleagues supplied. We grew crystals that diffracted out at the higher resolution.'

5.6 THE WHY AND WHEREFORE

Dr McPherson stated in an article in *Scientific American* in 1989 that he has been interested in the structure of macromolecules since the 1960s, when he was a graduate student at Perdue University. At that time, he helped solve the structure of the enzyme lactate dehydrogenase. His greatest interest has been in the effort to understand the properties of crystal growth in space, and the underlying physics. Concentrating on theoretical research, Dr McPherson explains, means that 'I am not making a directed effort to improve human health; only to improve the tools which may be used then to improve human health. The knowledge we have gained through the course of these experiments is going to help us improve the underlying technologies of X-ray crystallography and crystal growth which can be used, we think, to improve human health.'

In 1992 Dr McPherson flew a crystal growth experiment during the International Microgravity Laboratory Space Shuttle mission, using the liquid–liquid diffusion method in a German apparatus called Cryostat. He found that the satellite tobacco mosaic virus crystal produced dramatic results, which he described in 1992 as 'quite remarkable'. The crystals had no visible defects, and a novel octahedral form of the crystal was produced in space. The size of the crystal – more than 10 times that achieved on Earth – allowed a resolution so improved that it represented an increase of nearly 50% in the amount of useful data which the scientists could use to determine the crystal's structure.

Examining the results of the space experiments, Dr McPherson proposed that the mechanism for enhanced order and reduction of defects is not directly due to reduction in convective turbulence at the surface of the growing crystal, which initially had been hypothesised, 'but to the modification of the local concentration of nutrient in the immediate neighbourhood of the growing crystals. In the absence of gravity,' he concluded, 'there is no convective mixing of the solution, and nutrient transport is dominated entirely by diffusion, which for protein molecules is extremely slow.' While the diffusion of protein molecules to form the crystal is slow, the diffusion of impurities is even slower, resulting in a higher-quality crystal. Thus, he theorises, 'a fundamentally uncontrolled growth process in the presence of gravity becomes self-regulating in microgravity.'

In an article published in *Trends in Biotechnology* in June 1997, Dr McPherson elucidated the importance of understanding the underlying physics of the crystal growth process. Microgravity experiments began in the mid-1980s, he reports, and 'experiments since then have been dogged by persistent skepticism in a substantial segment of the crystallographic research community.' There were several reasons for doubt, including that 'there was no persuasive explanation of why the elimination of convection and sedimentation should necessarily produce more-ordered or larger macromolecular crystals... There were, at least initially, no accepted quantitative criteria for evaluating relative crystal quality. Many of the early experiments were impossible to reproduce and often poorly documented.' One serious problem was infrequent flight opportunities, so it was difficult to duplicate experiments, and there were 'excessive expectations, and overly enthusiastic appraisals based on qualitative data. There were relatively few experiments in space in the devices being used.'

With results available from the density of crystal growth experiments on Mir, 'a convincing theoretical framework now exists for understanding the phenomena involved. Physical methods such as interferometry and atomic force microscopy have revealed the unsuspected variety, structure and number of dislocations and defects inherent in macromolecular crystals' grown on Earth. These tools have allowed an analysis of conventional crystals which revealed the defects to be 'far more complex, extensive and dense, by several orders of magnitude, than those commonly associated with conventional small-molecule crystals.' The conclusion that is reached is that macromolecules are more sensitive to the conditions of their growth – particularly the degree of supersaturation at which they are grown. This sensitivity may account for the superior results obtained with some crystals grown in microgravity rather than on the ground.

This effect is produced, Dr McPherson states in a 1999 paper, by the creation of a quasi-stable depletion zone, or concentration gradient, around the growing crystal. With convective mixing in gravity, the crystal is constantly exposed to the full concentration of protein nutrient present in the bulk solvent, as well as impurities. Without the convective effects of gravity, the crystal is surrounded by a lower concentration of bulk solution and impurities as it grows. Through the slower transport of impurities to the crystal than the protein molecules, a self-purifying effect is observed, and the slower growth of the crystal itself appears to be a more optimal growth path. Using a ground-based version of equipment he is preparing for use on the International Space Station, McPherson and his team have been able to quantify the process of fluid transport during protein crystal formation.

With such increased understanding of the fundamentals of protein crystal growth, scientists will be able to better determine which crystal growth technology is best suited for which proteins, and take some of the guesswork out of the important research that the growing of crystals in space can provide in the fight against disease.

5.7 STRUCTURE-BASED DRUG DESIGN

The institution that has carried out the most extensive collaborative work in bringing together NASA and other scientists with the pharmaceutical industry is the Center for Macromolecular Crystallography, which in 1985 was established exactly for that purpose. The Center's current Director, Dr Larry DeLucas, has been a major spokesman for the promise of this application of space technology to human health, and explained the details in testimony before the House of Representatives on 22 June 1993: 'Diseases are often caused or influenced by proteins within our body that are no longer functioning properly, or by foreign proteins that have entered our body, such as those introduced when we are infected by viruses, bacteria and parasites. Drugs used to help fight various diseases work by binding to the protein that causes the disease. In so doing, the drug alters the biological function of this protein, usually inhibiting its harmful action within our body.'

Until the present time, the path for new drug development has been to experiment with various possible chemical combinations. But 'trial-and-error methods of drug development often produces side effects', DeLucas explained, and, of course, can take decades. It has taken a decade, for example, for the pharmaceutical industry to develop a new line of antibiotics, as announced in September 1999. 'Targeted drugs can interact more effectively with the target protein', and can also lessen side effects by not interacting with other proteins in our body. 'If the structure is known, drugs can be tailor-made that selectively bind to target sites.' This approach is known as 'structure-based design'. Such drugs, with the proper shape and distribution of atoms, can bind tightly within key sites of the protein and block its biological action.

In a more recent paper in 1999, Dr DeLucas outlines the general approach to structure-based drug design. First, a target protein is identified, based on its involvement in the biological pathway of a disease. Knowing the molecular structure and chemical properties of the target molecule, which in some cases can be obtained

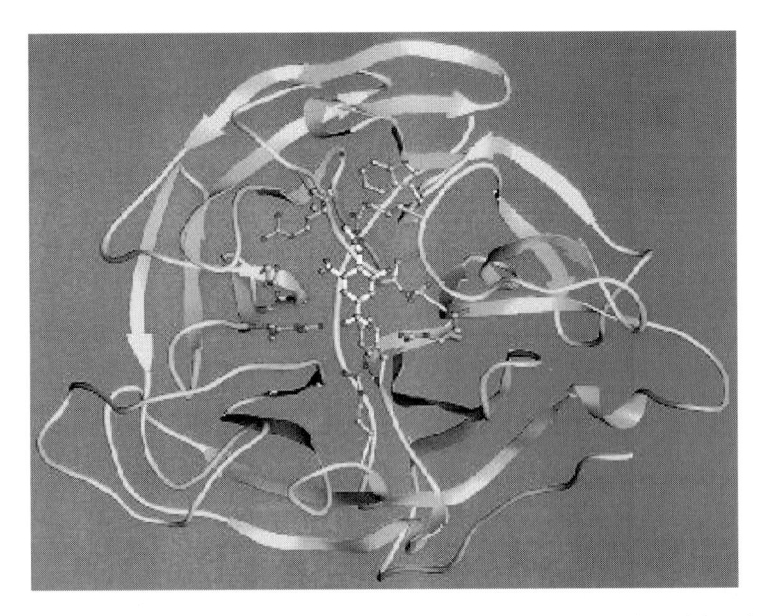

Figure 5.15. Structure-based drug design is being used to replace the trial-and-error approach to testing new pharmaceuticals. The Center for Macromolecular Crystallography has developed the Ribbons programme to produce computer-generated images such as this one, to aid in the design of new drugs. In this rendering, a small white molecule in the middle of part of the influenza virus effectively blocks the virus from reproducing itself.

with the best resolution by growing the protein crystals in space, scientists design compounds, atom by atom, to fit the shape and chemical reactivity of the target protein. They can also use computer programs to search available chemical data bases to select compounds that fit the target's active sites. By fitting small molecules into the active site of this protein, the enzyme produced by the protein is impaired, thereby halting or altering the disease process.

There is also the possibility of genetic protein engineering using recombinant DNA technologies. Currently, Dr DeLucas reports, investigators can modify DNA molecules that code for protein. This way of altering the functioning of the protein could lead to entirely new types of enzymes that can be used as catalysts for mediating a variety of reactions, and for the design of pharmaceutical agents that have carefully engineered physical and biological properties.

There is now an impressive list of protein crystals appropriate to medical applications that have been produced in space which are superior to anything produced on Earth. These include: gamma-interferon, important in antiviral research and in treating certain types of cancer; human serum albumin, the most abundant protein in blood, responsible for the distribution of many different drugs to various tissues; elastase, a key protein known to cause the destruction of lung tissue in patients suffering from emphysema; NAD synthetase, important in the development of broad spectrum antibiotics; isocitrate lyase, important in the development of anti-fungal drugs; proline isomerase, important in tissue rejection; and glyceraldehyde phosphate dehydrogenase, a protein important in the devastating parasitic disease, chagas.

Dr DeLucas stresses that research must continue, even on the crystals that have already been successfully grown in space. One problem is that the X-ray diffraction analysis method damages the protein in the course of the X-ray bombardment, and so a supply of crystals is required; and a constant supply of crystals is necessary for the drug design phase for various diseases.

While it might be tempting to limit the time taken to develop new pharmaceuticals, Dr DeLucas cautions that 'it is difficult to predict when a useful drug will result from protein crystals grown in space. One must realise that the development of pharmaceutical compounds typically takes anywhere from six to 20 years from the initial basic research stages all the way through the required clinical trials needed for FDA approval.'

But there are a number of very promising projects underway to develop new drugs, and other treatments and cures for diseases, that are enhanced by this science of protein crystal growth in microgravity. Working with the Center for Macromolecular Crystallography are 35 universities in Chile, France, Costa Rica, Germany, Japan, The Netherlands, Italy, Brazil, Great Britain, Canada and Israel, along with 24 partners in industry.

In a briefing for reporters at the Johnson Space Center in Houston on 12 May 1998, Dr DeLucas reported on market-viable drugs being developed from data provided by space-grown crystals. Drugs currently undergoing trials – and expected to be available in the next 2–6 years if they pass Food and Drug Administration tests – would treat bacterial and fungal infections, influenza and stroke.

One protein under intense study is Factor D, which plays a destructive role when there are complications from open heart surgery. It is an enzyme implicated in the production of inflammation after surgery, due to its involvement in the immune system. On the 1992 US Microgravity Laboratory-1 flight, on which Dr DeLucas was a crew-member, a crystal of Factor D that could be used for high-quality X-ray data was produced. Crystals grown on the ground had not been of the quality to produce high-resolution diffraction data. The space crystal allowed the completion of the three-dimensional structure of Factor D, Dr DeLucas explained at a conference in 1999. Within six months of the space-based experiment, a structure-based drug design team had identified a lead inhibitor compound. Subsequent research led to the development of other inhibitors, using a combination of Earth- and space-grown crystals, and the BioCryst company has recently begun human trials with one of these second-generation inhibitors.

Chagas is a crippling parasitic infection carried by the 'Kissing Bug', and so far it is immune to treatment. It is a tropical disease, prevalent in Latin America, which wastes muscle tissue until the patient dies, and is now found in 150,000 people in the United States. In July 1997, NASA flew the Microgravity Science Laboratory-1 (MSL-1) mission, the 35th flight in the space protein crystal growth programme. There were three different protein crystal growth experiments on MSL-1, with a record number of 1,500 protein crystal samples. Two of the proteins that were grown in the Vapour Diffusion Apparatus from the Center for Macromolecular Crystallography, are related to chagas. The hope is to find a compound that blocks a key protein in the disease. This an international effort in which the crystallography research is being carried out in collaboration with scientists in Argentina, Brazil, Chile, Costa Rica, Mexico and Uruguay.

Influenza, which kills up to 20,000 people annually in the United States alone, is also being studied for targeted drug design. Viruses have been described as 'bad news wrapped in a protein'. During a press conference while the Microgravity Science Laboratory-1 mission was in progress in July 1997, Dr DeLucas remarked that a virus is like a micro-machine that injects itself into a living cell, 'pirates the cell's chemicals to make copies of itself, then spills out more viruses to repeat the process.'

Scientists have located two proteins on the surface of the influenza virus that allow it to escape from the infected cell and attack other cells. While the virus mutates to change its appearance and fool the immune system, these two proteins remain unchanged. Dr DeLucas reported that scientists are using the knowledge they have gained from deciphering the influenza virus' coating to develop an inhibitor that prevents the spread of the virus to other cells. This would prevent the symptoms from worsening, and allow the immune system to mobilise to fight the disease.

Dr DeLucas reported that space-grown crystals of the protein coatings helped in deciphering the structure, and that the inhibitor being developed 'blocks the major strain of the flu'. While it would not cure influenza, he said, it would 'hold it at bay.'

5.8 DIABETES: A CASE STUDY

Diabetes – the inability of the body to regulate blood sugar due to ineffective production of insulin – often leads to serious complications, and is the third leading cause of death in the United States. Insulin – a hormone that regulates the metabolism of sugar – circulates in the bloodstream and interacts with all of the cells in the body. An estimated 5–10 million Americans suffer with Type 1 diabetes, in which a failure in the immune system directs the body to kill its own Beta cells in the pancreas, which sense sugar levels and produce insulin. (Type 2 diabetes often occurs in older and obese people, when the body develops a resistance to insulin). Together, Types 1 and 2 diabetes and their complications account for about $1 out of every $7 spent on health care in the United States, and a significant number of premature deaths.

On 8 June 1997 a Space Act Agreement was signed between NASA's Office of Life and Microgravity Sciences and Applications and the Juvenile Diabetes Foundation, to carry out joint research activities and to transfer technologies developed by NASA to researchers in the diabetics research community. The activities by NASA include protein crystal growth experiments, three-dimensional tissue culturing (discussed in the preceding chapter) and noninvasive diagnostic technologies for improved treatments.

Even before the formal agreement was signed, the Center for Macromolecular

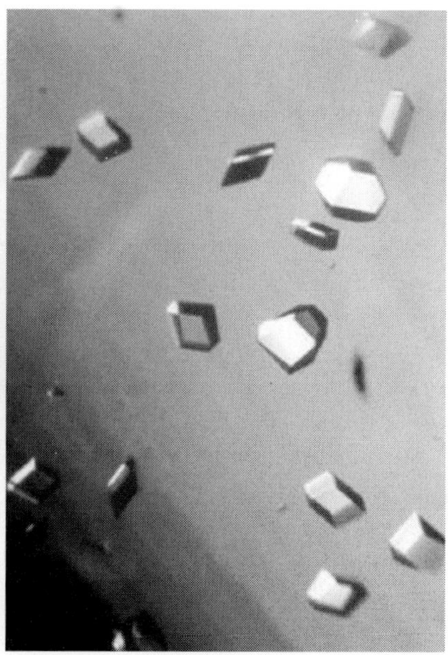

Figure 5.16. Diabetes is one of the most prevalent disabling diseases in the world. Crystals grown in space, seen here on the left, are larger and of greater clarity than those that can be grown on the ground. This image is of recombinant human insulin crystals grown during the STS-60 mission in February 1994. The Principal Investigator for the experiment was Dr Larry DeLucas.

Crystallography had been conducting research, along with prestigious researchers in the field, to find new ways of treating this disabling and deadly disease.

The Hauptman–Woodward Medical Research Institute of Buffalo, New York, is working with the Eli Lilly pharmaceutical company, using data from space crystals of human insulin, to design a drug that will allow the body to absorb insulin more effectively. The goal is to relieve diabetes patients from the burden of having multiple injections each day, to perhaps one injection every three days, and to smoothe out the 'highs' and 'lows' of blood sugar levels that diabetics suffer between injections. This would eliminate many of the debilitating complications of the disease, including kidney failure, blindness, and loss of fingers, toes, and even legs, due to the surplus sugar in the fine arteries and veins.

Working under grants from NASA and the National Institutes of Health, Dr G. David Smith, from Hauptman–Woodward, was the principal investigator of an insulin-growing crystal experiment during the Space Shuttle STS-60 mission in 1994, conducted through the Center for Macromolecular Crystallography in its Protein Crystal Growth Facility.

Insulin has two forms – monomer and hexamer. The monomer form is absorbed rapidly in the body, but is not stable, and so the body does not store insulin in that form. The inactive hexamer molecules, which are stable over time, gradually dissolve to deliver the active monomers into the bloodstream, to be absorbed. But insulin is

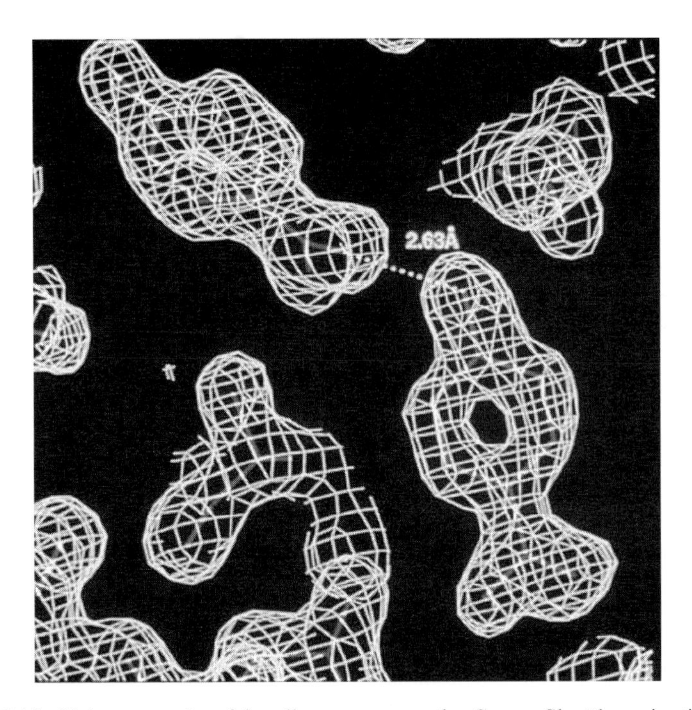

Figure 5.17. Using crystals of insulin grown on the Space Shuttle, scientists at the Hauptman–Woodward Medical Research Institute in Buffalo, New York, have been able to analyse the structure of this protein, creating molecule diagrams to advance their research into treating and curing diabetes.

so complex that even the hexamer form has three different configurations. One variation, T3R3, slows down the delivery rate to the bloodstream, and can provide a stable blood sugar level over a period of days.

Dr Smith explains that 'insulin has this phenomenal property called allosterism. It can change its shape spontaneously, yet still be the same molecule.' If this inactive T3R3 form of insulin were made available to a diabetic patient, Dr Smith explains, it would provide a diabetic with 'a longer basal level of insulin'. A single dose could provide a stable blood sugar level for three or four days. He said that there are a number of chemicals that can cause the T6 form of insulin to flip to T3R3, 'but we don't know why insulin wants to do this'. It is not a simple task to try to modify the hormone, Dr Smith stresses: 'It's an amazing molecule. In certain places the molecule is very forgiving to changes. But if you make too big a change, you wind up with a molecule that is inactive.'

On the STS-60 Space Shuttle mission in 1994, well-ordered insulin crystals were produced during eight days in space. Using X-ray crystallographic techniques, an exceptionally detailed map of the T3R3 hexamer form was produced. Dr Smith reports that 'the space-grown crystals have provided us with new, never-before-seen information. As a result, we now have a much more detailed picture of insulin.' Images of Earth-grown crystals resolve to features at 1.9 Å (one hundred millionth of a centimetre), but on STS-60, human insulin crystals provided 1.4 Å electron density maps. With this better resolution, scientists can see how structures attach in the crystal, and hope to design chemicals which will spontaneously generate T3R3 hexamers of insulin that are stable when stored, but dissolve at the proper rate for the diabetic patient.

In a report by NASA in July 1998, Dr DeLucas is optimistic, but cautions that at least three years of research and trials are required before the new knowledge can be translated into new diabetic therapies. The hope is that a more controlled release of insulin may ultimately decrease the medical complications from this prevalent disease.

The idea of engineering a crystal of an improved protein or hormone for direct injection into a patient, which is taking advantage of superior space-grown crystals in the development of improved diabetes treatment, is also being investigated for applications to diseases besides diabetes.

Interferon Alpha-2 is a protein pharmaceutical that is sold in more than 50 countries worldwide to treat as many as 14 different illnesses, including hairy cell leukemia, multiple myeloma, veneral wart, AIDS-related Karposi's sarcoma, and chronic hepatitis B and C. Taking their cue from the long history of injectable crystalline insulin doses by diabetics, scientists from Schering–Plough Research Institute are investigating the possibility of time-released injection of alpha-interferon in a crystalline form. When patients receive alpha-interferon today, its serum half-life is 2–6 hours with subcutaneous administration, and only minutes through intravenous injection. It also has a 'pulse profile', in which bursts and then rapid declines in the blood are measured.

The objective is to produce a time-controlled release formulation for alpha-interferon, so that there is an effective level over a significant period of time. Toward that end, space-grown crystals have been administered to primates, and the scientists observed that the serum half-life was twelve hours, compared to two to three hours in the non-crystalline form. There is the hope that a new drug delivery system will be developed through this space crystal research.

Dr DeLucas summarised the state of space-based protein crystal growth in his July 1997 press conference, NASA reported. In seven years of space-based research, he said, from crystals grown in experiments on 35 Space Shuttle missions, investigators have defined the structures of 30 proteins. The first few flights, he explained, were needed just to find the best conditions for growing protein crystals in this new environment. While it may seem that the public has read the same story about the same specimens being carried again and again on different flights, 'they had to be, as scientists focused on a different aspect of the structure and, in the process of X-raying the specimen, destroyed it,' he stated. 'To determine a crystal's structure, you need a continous supply, maybe even for an entire year, and even more to design a drug that works with it. We need a constant supply of crystals to determine the three-dimensional structure.'

The space protein crystal growth programme has been criticised, Dr DeLucas stated, because with the first-generation technologies, only 20% of the specimens flown succeeded in producing the best samples for X-ray diffraction analysis ever

Figure 5.18. Seeing the unseen is the purpose of growing protein crystals in microgravity. These crystals of turnip yellow mosaic virus were grown in the European Space Agency's Advanced Protein Crystal Facility, which flew on the Space Shuttle in July 1994. Dr McPherson and his colleagues report that these crystals display a unique, multifaceted form which has been observed only in space-grown crystals. The photograph is at x40 magnification. (Courtesy Dr Alexander McPherson.)

collected for that crystal. This is largely because scientists are still learning what the ideal growth conditions are which vary with each protein. Most researchers have also been limited to two weeks on the Space Shuttle, which is not long enough for many crystals, he explained.

Dr McPherson placed the space experiment results in their proper perspective in an interview: 'You've got to remember that when you're doing it on the ground, you've got a bunch of technicians standing around, you've got all your apparatus, you can go have a cup of coffee and do it at your leisure, you can try fifteen different things, you can use your own hands. When you do it in space, it's a completely automated procedure where you have six samples. You can't vary anything. The experiment is infinitely more constrained in space than it is in the ground. So it's not surprising that in any kind of marginal case, you're going to get better results on the ground than you get in space. It's no big shame if you do something in space and don't get as good a result as you do on the ground. The amazing thing is that you ever get anything *better* in space than you do on the ground, given the difficulties that are imposed on you there.'

Despite the obvious constraints that exist in space, and the verifiably superior results that have nevertheless been obtained in some experiments, on 9 July 1998 the American Society for Cell Biology released a statement reviewing NASA's life sciences programmes, incredibly concluding that 'no serious contributions to knowledge of protein structures or to drug discovery or design have yet been made in space.' Taking on the role of the Luddites of the scientific community, they conclude that 'there is no justification for a NASA protein crystallisation programme, and this committee strongly recommends that no further funds be spent on crystallisation of proteins in space.'

Keith Cowing, editor of *NASA Watch*, characterised this by stating that 'the authors of the ASCB report, with virtually no microgravity research expertise, hastily threw together a report designed specifically to validate their own personal opinions, using newspaper articles and press releases (but not any peer reviewed research) as references, put it all in the form of a two-page press release, and issued it in the name of the ASCB. This report has since been repudiated by the broader scientific community.'

The timing of the release of the report was not accidental. It coincided with the annual press conference by Representative Tim Roemer (Democrat, Indiana), calling for Congress to terminate the International Space Station, using the same flawed logic.

But the scientists who have seen the results from their experiments in space are not deterred. In an interview, Dr Dan Carter stated: 'The fact that you get an improvement [in space] is no longer controversial, as far as I'm concerned. You've got too many independent laboratories that analyse their own crystals, that are publishing papers that show the improvement. Now we're starting to understand the physical processes involved, that it's not snake oil or anything like that. There are physical reasons why this occurs, and we're beginning to prove those.'

5.9 PREPARING FOR THE INTERNATIONAL SPACE STATION

All of the teams of researchers who have developed protein crystal growth techniques for the Space Shuttle and Mir are now mobilised to design new, improved facilities for long-term experiments on the International Space Station. In a briefing at the Johnson Space Center in 1998, Dr Larry DeLucas pointed out that the 'key advantage of the ISS is the fact that we will be able to do science all year long. We won't have just a glimpse at it, as we do with the Space Shuttle. You will put your experiment up there, and for a long period of time we'll be able to do successive experiments, just as we do here on Earth. That's the only way to make progress in any scientific field.'

Because more than 40% of the proteins crystallised on Earth require more than three weeks for the process to be completed, with some extending to three months, the International Space Station will provide the only opportunity to study a large percentage of the more than 1,000 proteins that play a role in human health.

One of the areas that will be greatly advanced for research on the ISS is the on-orbit analysis of protein crystals as they grow, and when they have reached maturity. Dr DeLucas points out that space-grown crystals can suffer harm upon Earth re-entry, and that they can also deteriorate on orbit over time. There can also be a four- or five-day delay in returning the samples from the Space Shuttle to the investigator's laboratory, adding to the deterioration in the quality of the crystal.

Figure 5.19. The Observable Protein Crystal Growth Apparatus has been designed in Dr McPherson's laboratory to add to the theoretical base of knowledge about protein crystal growth in microgravity. It will fly on the International Space Station, and use time-lapse photography to visualise the fluid environment around the growing crystal. In this photograph, Joel Kearn, of the Marshall Space Flight Center, is viewing a laboratory demonstration of the facility.

While scientists will still hope to bring back protein crystals grown in space for analysis in laboratories on Earth, increasingly sophisticated diagnostic facilities on orbit will allow them to study both the crystals themselves, and the processes of their growth, directly in the environment of microgravity.

Dr Alexander McPherson and his colleagues hope to use their Observable Protein Crystal Growth Apparatus (OPCGA) onboard the ISS in order to add to the theoretical base of knowledge concerning the enhanced growth of protein crystals in space. The OPCGA will use the liquid–liquid diffusion method to produce crystals. The crystals will be observed using time-lapse video microscopy with polarised light, and interferometry to visualise the fluid environment around growing crystals. Ground-based tests with this new device have already allowed the scientists to record data, allowing a diagrammatic representation of depletion zones around crystals growing in the laboratory.

Before the ISS is ready to receive the full Observable Protein Crystal Growth Apparatus, Dr McPherson's cryogenic dewar protein crystal growth apparatus, based on the device successfully flown on Mir, will be deployed on the new space station. 'We're going to be the first scientific experiment flown on the ISS,' he reported in an interview, 'outside of the physiological measurements on the astronauts.' The experiment can go any time, 'as long as there is a temperature-controlled pressurised node, we can fly.' The dewar experiment does not have to wait for there to be a crew on the ISS.

The second-generation enhanced dewar for the ISS will have temperature recorders and a heater, he explained, to more precisely manage the crystal growth process. The heater is to control the rate that the samples thaw. Unlike the OPCGA, which will aim to produce a limited number of crystals of high quality and analytic usefulness, the dewar will carry many samples, and so candidate crystals can be screened in space.

The group of scientists at the University of California are also taking advantage of the requirements of their ISS dewar experiment, to involve students in a programme which they call 'ISS: A Science Classroom for America.' Dr Stan Koszelak explained in an interview that the need to prepare a large volume of samples to fly in the dewar (it can accomodate up to 10,000) presented an opportunity to involve young people both in their preparation and in the experiment itself.

Each sample that is readied to fly in space is prepared by hand, and then frozen by plunging it into liquid nitrogen. This is a relatively simple process, requiring limited manual skills and training. Dr McPherson and his team report in a 1999 paper that 'the process is so simple, it occurred to us that high school students could directly participate in flight experiments aboard the ISS. Students, we believe, could be enlisted to make protein crystallisation samples for the Enhanced Dewar Program, under strict supervision and guidance by NASA approved, trained, and experienced personnel, and these samples [could be] frozen and carried into space.'

The plan is to return the samples that fly in space to the classrooms and students who prepared them for examination, and to record the results. 'The experiments would, therefore, provide both a student flight experience, but at the same time also be part of an important, ongoing scientific investigation.'

Because only a limited number of students can participate directly in the space experiments, and children in lower grades may not have the ability to participate, a classroom experiment kit has been created which includes all of the materials that the students need to mix the protein solutions and to screen the resulting crystals in school.

Dr Koszelak reported that some of the high school children participating in the flight experiment will be invited to visit either Dr McPherson's laboratory at the university or NASA's Marshall Space Flight Center, to produce the samples that will fly in space. So far, about 30 students have been to the university, and between 50 and 60 have been to Marshall. The scientists hope to use this experience with spaceflight as an inducement for the students to learn structural biology.

5.10 X-RAY DIFFRACTION EXPERIMENTS ON ORBIT

The scientists working with Dr DeLucas at the Center for Macromolecular Crystallography are also preparing for the ISS era. They have developed an X-Ray Crystallography Facility (XCF), which will allow scientists to examine crystals in orbit, select some to be preserved and returned to Earth, and carry out *in situ* diffraction analysis of a small number of them. The XCF is described as a 'full crystal analysis laboratory', where crystals will first be harvested from their growth chambers. Up to 200 crystals can be snap frozen, and stored for transfer to the Space Shuttle for return to the ground at the end of each 90-day ISS astronaut increment. 60 of these 200 will be selected through a partial diffraction analysis, which requires 30 minutes for each crystal. The remaining 140 crystals that can be preserved will be selected based on video examinations by researchers on the ground, but between 10 and 16 crystals will be chosen for a full diffraction analysis onboard the ISS. The data acquired through the on-orbit X-ray diffraction will be downlinked to the crystallographers on the ground, who can then plan the next steps in their research without having to wait for the samples to be returned to Earth. The diffracted crystals will be discarded on orbit after the useful data has been transmitted to the ground.

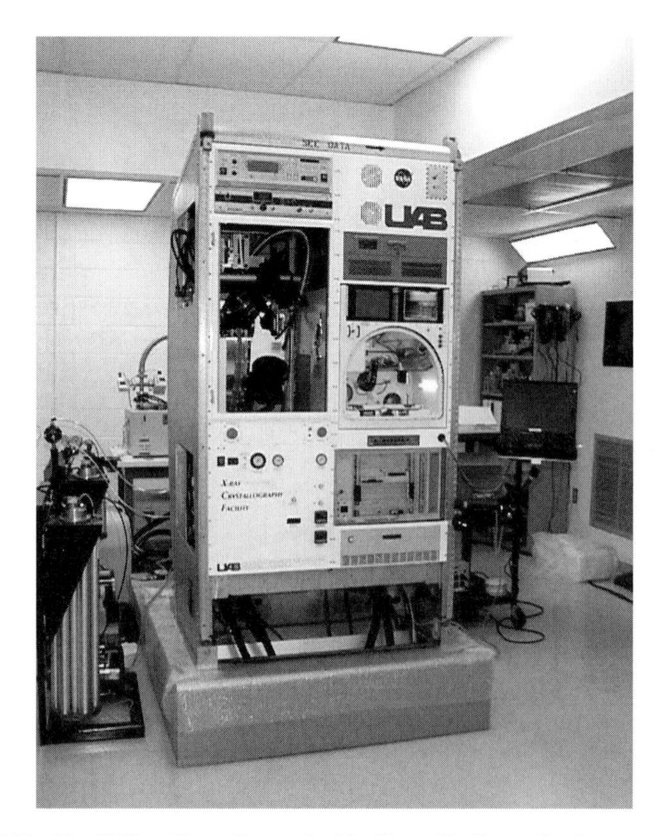

Figure 5.20. The X-Ray Crystallography Facility will allow scientists to perform X-ray diffraction on candidate crystals grown on the station without having to subject them to the return to Earth. In this photograph, a prototype of the XCF is installed in a Standard Payload Rack, as will be the flight hardware on the International Space Station. (Courtesy Dr William McDonald/Center for Macromolecular Crystallography.)

In an interview, developer William McDonald explained that the instrument will be controlled from the ground. The crew will be needed only if a problem arises. It will be deployed in 'bits and pieces', he said. The crystal growth facility, which uses the vapour diffusion method, will fly first, and the full X-ray diffraction facility will follow in steps. It has been designed for a five-year lifetime.

Before this equipment could be qualified to fly to the International Space Station, it was necessary to determine if the natural radiation environment in space would interfere with the X-ray diffraction that the scientists hope to use in orbit. During the last astronaut increment on Mir, with Dr Andy Thomas, the DeLucas group flew the X-Ray Detector Experiment in order to examine the radiation environment inside the spacecraft. Studies have been carried out to analyse the radiation *outside* spacecraft, but not inside.

It was necessary for this experiment to be undertaken on Mir, because it was in the same approximate orbital inclination as the ISS will be – about 57°, as compared to the Space Shuttle, which is inclined 28° from the equator when it is not flying to the station.

The radiation inside a spacecraft derives from solar protons and background cosmic rays. The concern was not that the radiation would damage the crystals, but that it could produce 'false positive' signals in the X-ray detector. McDonald explains that if a high-energy particle hit the detector 'it could produce a flash of light that would be picked up, and could be interpreted as a diffraction point in the crystal.' He stated that signal processing techniques could remove a small number of such superfluous spots, but that 'the image would be contaminated with a large number.'

McDonald reports that the background radiation onboard Mir was found to be 'quite low' in the 'quiet' part of Mir's orbit, where there were between three and six

Figure 5.21. The International Space Station will allow hands-on interaction between the crew-members and the scientific experiments on board. Astronauts and scientists, such as Dr David Wolf, seen here on Mir, will be able to be the extended eyes and brains of the scientists on the ground. In this photograph Dr Wolf is operating the Interferometer Protein Crystal Growth Apparatus on Mir, which allows him to observe how the concentrations vary inside a solution as a protein crystal grows in space.

high-energy particle events per square centimetre per second. But when the spacecraft passed over the South Atlantic Anomaly, where there is a dip in the Earth's magnetic field, the rate increased to about 60 hits. The dip allows cosmic rays and charged particles to reach lower into the atmosphere. During its mission, the Röntgen Satellite, an X-ray observatory, turned off its X-ray detector during passes over the South Atlantic Anomaly to avoid damage to the equipment. The team is investigating methods of removing any defects from the diffraction images created by the radiation. They can also shut the instrument down for the 3–5 minutes that the ISS is over the Anomoly, which is only during three or four orbits per day.

On 8 March 1999, NASA announced the selection of 48 new biotechnology experiments to be developed for the Space Shuttle and the ISS, focusing on protein crystal growth and cell science. One of the scientists is Dr Robert Snyder, with New Century Pharmaceuticals. He is developing a system for purifying proteins and cells using electrophoresis, or passing an electrical field through a fluid as it flows from one end of a chamber to another. His goal is to produce purer proteins for the crystal growth community so that they can eliminate impurities that cause defects in crystal structure.

Dr Robert Naumann – who was also formerly at the Marshall Space Flight Center, and is now at the University of Alabama in Huntsville – is working on a method of restricting the diffusion of molecules to the crystal. This new crystal growth technique will limit the access which the molecules have to the growing crystal, which will mean it will take longer to grow them; but the result expected is an improved quality of the crystals, both on Earth and in space.

Other researchers will be examining the crystals on the genetic level, and new techniques to optimise the growth of protein crystals in microgravity. This work will create new avenues for next-generation research on the ISS.

For the first phase of protein crystal growth experiments and facilities on the ISS, the equipment will be operated with as little crew-time required as possible. Similar to the range of bioreactor experiments onboard the ISS, there will be both basic science and also applications, and commercially-oriented payloads. The goal is to advance the theoretical understanding of protein crystal growth in microgravity, as well as to continue the efforts to understand, treat and cure disease.

Dr Snyder, and others in the ISS community, hope that when the station is fully operational there will be the ability for scientists themselves to go into space to operate their experiments. 'There is no reason you can't have scientists on the ISS,' Dr Snyder believes. While a high degree of automation is necessary while the station is being assembled, 'the unexpected phenomena will be missed unless we identify and take advantage of a unique result.'

The ISS holds the promise of allowing scientists the opportunity to broaden the frontiers of science in understanding and improving human health, through research in fields such as bioreactor tissue engineering and protein crystal growth, and to lay the basis for the long-term exploration of space through studies of changes in human physiology, how to grow food in space, and areas of research that today can only be imagined.

6

The lessons learned from Mir

During the four and a half years of the Shuttle–Mir programme, US astronauts spent more than 975 days on Mir, which was more than the time spent on the Space Shuttle during 17 years of operation. They underwent training at the Gagarin Cosmonaut Training Centre, learning the fundamentals of 40 years of Soviet and Russian spaceflight, a new language and a different culture. Nine Russian cosmonauts flew on the Space Shuttle, training at the Johnson Space Center, undergoing a similar acculturation in both the technical and social arenas.

Over the course of the programme, American and Russian programme administrators had to learn to cope with the budgetary and political vagaries of both nations. Programme managers faced quiet times, as well as accidents and equipment failures on Mir, and delays in Shuttle flights to service the space station. The astronaut and cosmonaut crews, the ground support personnel in both countries, and the management from both the US and Russian space agencies had to develop trust in each other's capabilities and abilities to deal with abnormal circumstances.

While many concrete engineering lessons were certainly learned through the Russian–American cooperation on Mir, Shuttle–Mir programme director and astronaut Frank Culbertson believes that the most important lesson that NASA learned was 'humility'. For 40 years the space agency has directed a programme without any advice from 'outsiders'. But, 'as good as we are,' Culbertson said, 'we learned things we did not know. Just setting the bit in your mind that you don't have all the answers, as an agency or a country – that someone else has something to offer – that was a great leap forward.'

In the extensive 'Phase 1 Program Joint Report' issued by NASA and the Russian Space Agency in January 1999, the concluding chapter, written by Phase 1 Programme Managers Valeri Ryumin and Frank Culbertson, states: 'The Phase 1 programme endured through a fire, a collision, several power shortages, and other significant contingencies and last-minute adjustments, and proudly accomplished its four main objectives... US and Russian space programmes bridged cultural, linguistic and technical differences and created a joint process for analysis, mission safety assessment, and certification of flight readiness. This collaboration resulted in a joint programme spanning more than four years that capitalised on a combined four decades of spacefaring expertise in both Earth orbital and intercosmos exploration, to build the foundation for an International Space Station.'

The first chapter of the same report simply states that the 'largest benefit of the Phase 1 Program was the growth of trust and understanding between National Aeronautics and Space Administration and the Russian Space Agency.'

Precisely because the two space programmes had to deal together with unforeseen situations and several emergencies, through which trust developed, NASA Administrator Dan Goldin has stated that the Shuttle–Mir programme 'has proven to be an unprecedented learning opportunity for living, working and conducting research in space.'

6.1 REDUCING THE RISK

The primary goal of the joint effort was to reduce the risk or probability of failure, associated with developing and deploying the International Space Station. Although the United States has docked spacecraft in orbit, it has never assembled a multi-module facility, requiring dozens, if not hundreds, of hours of extravehicular activity. Astronauts have lived in space in the Space Shuttle, but not for more than two weeks at a time. American engineers and crew-members had worked briefly with the Soviet Union in the 1970s, to carry out the Apollo–Soyuz Test Project, but that was a one-time 'handshake in space', not seen as a long-term commitment for joint work. The Shuttle–Mir programme was designed to address these issues and provide the US with greater confidence in undertaking the International Space Station project, in which the Russians will be a major partner and science will be the major focus of activity.

NASA knew in 1993, when the US–Russian space agreement was signed, that the International Space Station would be the most difficult and risky effort ever undertaken in science, technology, engineering and international cooperation. NASA was also well ware that the Russians had nearly 20 years of experience on space stations, and wisely concluded that if that experience could be shared between the ISS partners, the risks associated with building, launching, assembling, using and supplying the new international station would be reduced.

In 1998, NASA's Office of Life and Microgravity Sciences and Applications prepared a report entitled, 'The Phase I Program: The United States Prepares for the International Space Station'. The report helps summarise what was accomplished during the first phase of the US–Russian joint space effort, the Shuttle–Mir programme.

One of the benefits to American space experience from the programme on Mir was simply the access to long periods of time in space, for both men and machines. One of the preliminary results from measurements taken from Mir, for example, indicates that the radiation environment around Mir measured over time, in an orbit where the ISS will be, is higher during active solar periods, and lower during periods of low solar activity, than NASA had previously believed. This more precise picture

Figure 6.1. The greatest benefit of the Shuttle–Mir programme, the two space agencies state, was the growth of trust and understanding between NASA and the Russian Space Agency. The crew patch for the STS-79 mission symbolises that cooperation. It was the fourth docking mission of the Shuttle to Mir, and the first to exchange American crew-members. As can be seen in the patch, Dr Shannon Lucid was returning to the Earth on the Shuttle, and John Blaha was going up to Mir.

of the amount of radiation that periodically 'leaks' through space from the Earth's Van Allen radiation belts, down to the altitude where the astronauts will be, is of more than just academic interest.

Space-walks (EVAs) are an integral part of the assembly and deployment of the ISS. Astronauts and cosmonauts will spend hundreds of hours outside the station to connect individual modules, place electric power-producing solar arrays, support deployment of the truss (structural spine) of the station, and mount and retrieve experiments on the outside of the spacecraft. As time goes on, space-walkers will also be called upon to make repairs.

As mentioned in the preceeding chapter, measurements of potentially damaging high-energy radiation inside Mir were also taken during an experiment prior to the deployment of an X-ray crystallography facility on the ISS. It was learned from such measurements that the South Atlantic Anomaly has moved 6° west and 2° north since it was previously measured in the 1970s during the Skylab programme, reflecting a dynamic process in the structure of the Earth's magnetic field. Knowledge of this increased radiation during certain parts of the ISS orbit will also be utilised to minimise the exposure to astronauts, radiation-sensitive equipment, and scientific experiments, during EVA.

With the understanding that the radiation environment is different from what scientists had believed, engineering and operational decisions are being made to minimise the astronauts' exposure to radiation during space-walks, by scheduling

Figure 6.2. One of the important lessons learned during the Shuttle–Mir programme was a more precise characterisation of the radiation environment in space. This has led to changes in NASA's plans for extravehicular activity (EVA – space-walks) during the assembly phase, and later servicing and repair activities of crew-members. Here, astronauts prepare to deploy the Mir Environmental Effects Payload on the outside of the station in March 1996.

them to coincide with quiet periods of solar activity. Engineering changes have also been incorporated to eliminate as many hours of EVA as possible.

Other monitors placed outside Mir found that its surface was being contaminated with a residue from propellant used in firing its attitude control engines. One such monitoring equipment package was the Mir Environmental Effects Payload, which was placed on the outside of Mir's docking module by Space Shuttle astronauts in March 1996. It remained outside Mir for 18 months, and was returned to Earth for analysis in October 1997. Using the new data, ISS engineers are changing the propellant venting procedures that had been designed for the new station, in order to avoid contamination.

The Space Shuttle also provided the Russians with an analytical capability which they did not have, for studying the long-term effects of the space environment on the structure and materials of Mir. Under the Shuttle–Mir programme, Mir crews installed two new solar arrays to provide electric power to the station, and retrieved a portion of one of the original arrays, for transport back to Earth via the Shuttle.

Analysis revealed that the solar arrays had suffered significantly more damage than had been anticipated, from the elimination of waste from Mir and from the firing of the Space Shuttle's thrusters to manoeuvre near the station. Based on this information NASA is modifying the waste elimination procedures for the ISS, and will also effect changes in the protocol for Space Shuttle operations in the proximity of the new station. In addition, based on the *in situ* data of how materials deteriorate on orbit, the space agency is factoring this information into the maintenance and replacement schedules for the ISS solar arrays.

One of the most challenging feats required to assemble a multi-module spacecraft that will be larger than a football field, is to anticipate oscillations in its flexible structure when any force is applied to it on orbit, such as the docking of a Space Shuttle, Soyuz or other transport vehicle. Because it would collapse under its own weight in the gravity environment on Earth, no full-scale model of the ISS can be tested for structural stability before it is deployed on orbit.

To test the accuracy of the theoretical calculations of how the flexible ISS would behave under stress in space, NASA scientists developed precise mathematical and engineering models in an attempt to predict the behaviour of the Shuttle when it was docked to Mir. When the two spacecraft were docked, the Shuttle used its own thrusters to affect Mir's orientation, as the station was put into a passive, free drift mode. The engineers were gratified to find that their equations accurately predicted the dynamic behaviour of the combined Shuttle–Mir stack, which has validated their model and provided them with more confidence in their calculations of how the ISS will react to external forces. The NASA–RSA Joint Report, evaluating the lessons learned from the Shuttle–Mir Phase 1 programme, states that the successful Shuttle attitude control of the mated Shuttle–Mir stack – a total of more than 200 tons – 'was a significant milestone in the Shuttle programme'.

There was no lack of exciting moments during the Shuttle–Mir programme, as astronauts and cosmonauts had to battle an onboard fire, work in the dark when power was lost, re-establish control over the station when attitude control gyrodynes shut down, and seal off the depressurised Spektr module after the collision before the station's entire atmosphere vented out into space.

Anyone who believes that a fire, similar to that which took place on Mir, could never happen on the International Space Station, is a Pollyanna, not an engineer. One important lesson learned from the February 1997 fire was that the temporary shut-down of the station's ventilation system can prevent a fire from spreading. NASA has re-evaluated fire control procedures for the new station, and modified ISS software so that a single-command ventilation shut-off could be temporarily executed between modules in case of emergency. In addition, the location of critical hardware, such as fire extinguishers and medical kits, will be quickly and easily accessible to crew-members.

In June 1999 the depressurisation of the Spektr module after its collision with an unmanned Progress supply ship demonstrated the importance of maintaining clear and uncluttered passageways between modules. The lessons drawn from the Spektr damage also led to the redesign of some of the critical ISS components that are being made by the Russians, which will be more robust in the event of a depressurisation in one part of the ISS. It has also been determined that the crew should have portable life-support sensors to monitor the pressure, oxygen content and other parameters of

Figure 6.3. NASA's Crew Return Vehicle, seen here during drop tests from a USAF B-52, is being developed to allow all seven crew-members to return to Earth in the case of an emergency on the International Space Station. This free-flight test was successfully completed on 9 July 1999. (Courtesy NASA/Dryden, Carla Thomas.)

the atmosphere on board. This would alert them to any potentially dangerous leak, or other changes in the air upon which they depend.

Although engineers and designers must try to consider everything that can go wrong on a spacecraft, in order to plan to effectively mitigate their consequences, actually having to face failures and emergencies forces more detailed consideration of how to reduce potential hazards. During the initial planning it had been decided that once there was a full complement of seven crewmen on orbit, there would be two small Crew Return Vehicles, each able to accomodate 3–4 crew-members, attached to the ISS at all times. Now, other possibilities are being considered, such as a larger Crew Return Vehicle that can bring all seven crew-members back to Earth in a single vehicle, if there is a life-threatening emergency, along with a Russian Soyuz crew vehicle attached to the station at the same time, in order for there to be access to an emergency return vehicle at all times, by all members of the crew, no matter where they are in the station. Keeping a small Soyuz at the station, along with the larger Crew Return Vehicle, also allows for one or two crewmen to return in the case of an injury or other health concern, while still leaving a return vehicle at the station.

A major lesson learned by NASA during its joint operations on Mir was the difference between operating a spacecraft like the Space Shuttle, that can be brought back to Earth quickly in emergencies for repair, and a spacecraft that will never be brought back, but must be maintained and repaired in space. The American space agency – and the American Congress and public – are not accustomed to anything being repaired during a Space Shuttle mission.

There is no room to store many spare parts on the Shuttle, and no mission sufficiently extensive to change out major components. To increase the safety and reliability of the Shuttle system, NASA has incorporated redundant sets of major systems, such as the computers, into each Shuttle orbiter. If a major system suffers failures on orbit, the Space Shuttle is returned to Earth for repair. Critical components are changed out on the ground as they approach their estimated lifetime, during routine maintenance, before they fail.

By observing the Russians, NASA learned that on a space station, non-critical subsystems can be operated until they fail, and can be changed out or overhauled when necessary. There is ample capability to have the tools available on the station to maintain and repair the spacecraft with the required supply of spare parts for critical systems.

The failure of a critical system onboard Mir – such as one piece of equipment for supplying the atmosphere with oxygen – created fits of hysteria in the American media, and therefore in Congress. However, the Russians have developed methods of providing back-up systems for critical functions using *different* techniques than are used in the primary system. This provides multitiered and robust redundancy for the most important equipment aboard the station. Initial plans for the ISS called for the provision of duplicate back-up systems in the event of failures. Mission planners

Figure 6.4. One of the most important lessons learned by NASA during the Shuttle–Mir programme was the difference between repairing a Space Shuttle orbiter when it returns to Earth, and having to rely entirely on in-orbit maintenance and repair of a space station. Here, cosmonauts Pavel Vinogradov (left) and Anatoly Solovyov install a new carbon dioxide removal system in the base block of Mir. Dr David Wolf took the photograph on 10 February 1998.

have now decided to follow the Russian model and use a range of different technologies as back-ups.

NASA also learned that the Space Shuttle must function as a transportation system, first and foremost, when it changes roles from being an independent spacecraft to servicing the International Space Station. In the past, scientists and mission managers spent years preparing scientific and technology payloads to fly on the Shuttle. With the orbiter being depended upon to bring a new crew and vital equipment to Mir, NASA found itself having to make virtually last-minute changes in the payload manifest, as the failure of some hardware on Mir required that a spare be quickly added in with the rest of the Shuttle cargo. There was great satisfaction that this highly complex spacecraft was able to adjust to a 'just-in-time' mode of operation.

6.2 DOING SCIENCE? ASK AN ASTRONAUT

A second goal of the Shuttle–Mir programme was to conduct peer-reviewed 'precursor' research on Mir, in preparation for the ISS. Here, the primary purpose was not necessarily to produce dramatic new scientific results, but to be able to test equipment and operating procedures for periods of time considerably longer than the two weeks on the Space Shuttle, to determine how they withstood the rigours of long-term space operation. To support the American science programme on Mir, the Russians agreed to allocate considerable resources to accomodate the US cargo being delivered to the station – a power requirement of up to 2 kW average per day, 70% of the astronaut's time, and 30% of the cosmonaut's duty time.

Including the areas of research discussed in detail in the three preceding chapters – growing food in space, tissue engineering using the bioreactor, and protein crystal growth in microgravity – the science programme carried out during the Shuttle–Mir programme consisted of about 150 peer-reviewed investigations that spanned virtually every major discipline of science, including fluid and combustion physics, materials science, and Earth remote sensing (see Appendix 3). Many of the scientists who prepared experiments for Mir had done so for the Space Shuttle. Mir gave them the opportunity to glimpse what the possibilities for breakthroughs, and also the possibilities of equipment breakdowns, will be when a full-time laboratory is available.

Important lessons on how to best conduct scientific research in an Earth-like laboratory in space – the major purpose of the ISS – were learned during the Shuttle–Mir programme. These lessons have had an impact on how NASA trains astronauts, how experiments are selected and will be carried out in space, and how the astronaut–researcher's time should be organised to optimise the science results.

Dr Shannon Lucid was fortunate that her 188-day stay in orbit took place before the trauma of the fire, and before the disruptions caused by the Progress collision onboard Mir. A biochemist by training, Dr Lucid looked forward to 'going back to the laboratory' to carry out research during her stay on Mir, for which she had not had the opportunity since becoming an astronaut.

Out of necessity, Dr Lucid was left to her own devices in conducting much of her research on Mir, because there was only very limited communication with NASA ground support personnel in Moscow, and even less with principal investigators whose experiments she was trying to operate. 'Some people look at that [limited communication] as a detriment,' she said in an interview, 'but if you think about it, it can really work to your advantage. Actually, it worked to the advantage of both the person onboard Mir, and the person on the ground. The reason I say that,' she explained, 'is, at first, you could only talk to the ground [from Mir] for maybe ten minutes a day. And a lot of times the comm[unication quality] wasn't very good. That forced you into working in a mode that was very different than you work on a Spacelab mission on the Shuttle. Obviously, if I ran into a problem first thing in the morning, I couldn't sit there and wait until late in the evening when I talked to the ground [to solve it], or nothing would ever get done. So that forced me to really think through what was going on, and to try to be a little creative. You had to rely on yourself to come up with some answers. So many times on Spacelab flights, [with] the first little thing that is a little bit different, you immediately call down to the ground. You have this whole host of people sitting down there, and they're just basically there to help you.' That may work during a two-week mission, she indicated, but when a scientist has an experiment operating on a space station for months, he or she cannot always be available to immediately answer an astronaut's questions.

Referring to her more independent situation on Mir, Dr Lucid said 'I think it worked to the ground's advantage, too. They were there in Russia. They had so much to do... If someone had to be on call all the time to just answer my questions, they wouldn't have been able to get all the paperwork and the negotiations done that were necessary.'

Dr Lucid reports that 'whenever I did talk to the ground, I thought a little ahead about what I was going to say, and rated things, and said to myself, 'This is the most important. If the comm quits, I will have at least have got that question in.' It made me think about the quesitons I would ask the ground, so I didn't sent them off on a wild goose chase, which sometimes happens on the Shuttle. If you say something – the first thing that pops into your head – and you haven't really thought through what you're asking, people can spend a lot of time working on that, but if you'd taken a moment to think, you would have asked in a different way, and come up with a different answer.'

One of the operational areas that has created friction between the astronaut trying to carry out research and the ground personnel and NASA management, from Skylab through the Shuttle programme, has been the demands of a timeline of activities made on the crew-member from Mission Control, and the attempt to 'micro-manage' every minute of the astronaut's time.

I asked Dr Lucid how much flexibility an astronaut should be given in setting and meeting research goals during long-duration station missions, and what we learned in this regard from Mir. She explained that Mission Control put together a plan for the crew activities, which was sent up daily to Mir. 'But in reality,' she said, 'it just wouldn't work that way. The ground might say 'at 10 o'clock, you start working the furnace.' In reality, they don't know what's going on onboard Mir, so the way I

Figure 6.5. Crew-members on a space station have to rely largely upon themselves, Dr Shannon Lucid reports, and had to solve problems with experiments on Mir to a greater degree than during Space Shuttle flights, where they can rely more on the ground. Dr Lucid is seen here next to the microgravity glove box which she used on the Priroda module on Mir. In the centre is Flight Engineer Yuri Usachov, and on the right is Commander Yuri Onufriyenko.

used the flight plan was that I looked at it and said to myself, 'OK. Today I have permission to do this experiment, this experiment, and this experiment. So I'll start working on them, and when I get done, I get done, and I'll let the ground know.' I didn't get wrapped up with the fact that at 10 o'clock they expect me to do this. It didn't make any difference to the ground when I did it, just so long as it got done. And I'd talk to them about it.'

Dr Lucid is not alone in recommending that an ISS expedition crew, which will be in the space laboratory for at least three months, be given the latitude to set their own schedule to accomplish their work. 'That's actually a big thing that all the people from Mir have come back [and said],' she reported. 'They're trying to get that implemented on the International Space Station, so the crew can be somewhat autonomous in how they run their day, just like you do when you are working in a laboratory down here. That's what makes it so much fun. People enjoy the autonomy, so to speak, as to how they're running their day.'

Mir astronaut David Wolf – a medical doctor, researcher and engineer by training – reported in an interview that, like Dr Lucid, he was happy to return to work in a laboratory on Mir. 'Most of us came from working in laboratories,' he remarked, 'and once you're an astronaut, it's a different environment. I did feel exactly what Shannon felt, that it was a real pleasure to get back in the lab and do my work.' Dr

Wolf's science research programme was seriously effected by the hours of repair work that were needed in the ageing station, and he said that a major question for the future is: 'How do we schedule people and let them work?'

He provided an example from his increment onboard Mir: 'There was a time a month and a half into the mission, when I knew that to ideally image what I was seeing in the tissue culture, I needed to set up another whole instrument from another experiment – a video microscope – but that would have taken a few days, just to get it out and set it up, and there wasn't room at the time. I would have liked to call down [to Mission Control] and say, 'Look, we need to change course here, and take a few days out to document what we've got here [in the experiment].' But we couldn't do that, because we had other experiments coming on line. We had a schedule to keep.'

Dr Wolf reported that a key part of his debriefing, when he returned to Earth, was to say 'we need to have the flexibility. Very few discoveries come on schedule. We need to let discoveries happen. We need to have a flexible schedule, and optimal use of the human that is up there making judgements and observations.'

Dr Lucid is well aware, of course, that close coordination will also be needed with the ground. 'There is a certain amount of interaction that needs to go on between the crew person and the principal investigator,' for example. 'If I'm an investigator, and I have the candle experiment that is up on station, I need to know for my own personal planning that it is going to be running [at some specific time] so I can talk to the crew and make sure everything is going all right. I'm not going to go to the Carribean for my vacation during that time. I want to be available so Mission Control can contact me and let me know how my experiment is going. Then I can interact with the crew and answer their questions.'

To illustrate this point, Dr Lucid provided an example from her work on Mir: 'When I started up the candle experiment, that was a new experiment. Bill Gerstenmaier [in missions operations] had contacted the people at NASA's Lewis Research Center and told them 'Shannon will be starting on the candle experiments on this day.' This is in the middle of the night back in the United States, and he had them tagged in [on the line during] the one communication contact we were going to have that day, so they would be able to listen. If I came down with any questions, they would immediately be able to give Bill the answer [to give to me] so he wouldn't have to waste a lot of time having to contact them. They could listen directly to what I was saying.'

'It just so happened,' she continued, 'that the igniter was not working. I had worked most of the day trying to do it, but it flat didn't work. Then we got a comm pass, and Bill said, 'How's it going?' I said, 'Well, Bill, the igniter doesn't work,' and I described what I had done. Immediately, the person from Lewis said, 'That's because in zero g, it must not be making contact.' They told me one simple thing to do, I did it, and it worked like a charm from then on. That was a tremendous example of working together.'

In addition, there will be necessary coordination with the ground to be able, at any time, to execute a mix of experiments that match the resources and capabilities of the ISS. 'Power will be limited on the space station,' Dr Lucid explained, 'so if you had experiments that had big power requirements, you can't run two or three of them at the same time. You need to decide to do one experiment, and then start the next one. These kinds of things need to be taken into consideration, but there needs to be a tremendous amount of latitude for the crews.'

NASA management has indicated that it is taking seriously the recommendations of the Mir astronauts in planning the training and operations for science on ISS. In its 1998 Shuttle–Mir summary report, it is stated that 'Phase 1 lessons have emphasised that astronaut training objectives for long-duration crew-members will differ from those NASA has traditionally employed for Shuttle crews.' Russian training is skill-oriented and not matched to a specific task, because over a period of many months there is little possibility that a crew-member could remember all of the details of a particular experiment.

As Shuttle–Mir manager, and astronaut, Frank Culbertson pointed out in Congressional testimony in 1996, concerning the lessons already learned from the programme: 'Pre-mission training of crew-members, while it must still result in crew-members able to carry out the mission and the scientific experiments, must also take into account that it may be weeks or even months before certain activities trained for on the ground will actually be performed in orbit. The capability to retrain, or at least provide refresher training, during the mission must be built in. Preflight planning must take into account that it is virtually impossible to predict at the start of the mission what exactly will happen in the second or third month of the flight, and this flexibility must be built in.'

When the time arrives for a crew-member to perform a function which he has learned, in general, weeks or months before, aids will be needed. The NASA report states that 'feedback from early Phase 1 astronauts led mission planners to expand the crew orbit support of training, entertainment, language teacher, and information resource.' Astronauts who were onboard Mir after Shannon Lucid's increment, used a laptop computer to 'brush up on the particulars of experiments immediately before start-up, or to review EVA hardware and procedures before a space-walk.'

Echoing Shannon Lucid's concerns, the report states that the Phase 1 experience 'has taught us that it is neither practical nor feasible to create extremely detailed day-to-day timelines for long-duration space station operations. The Russian programme uses a more flexible approach to scheduling, in which crew-members apply the fundamental skills they learned in training to the tasks required by the actual priorities of the day.' The report cites examples of when detailed planning from the ground will not work. 'For instance, solving a problem with an experiment's equipment may require sending a replacement part or repair tool to the station on an interim flight. Meanwhile, rearranging the research agenda would free time to complete the problematic experiment once the repairs are complete.'

While close coordination with the ground is critical for mission success, I asked Dr Lucid if there could be too much contact between the crew and the ground. 'Frankly, when you're talking to the ground, you're not getting any work done,' she said. She compared her relatively quiet and uninterrupted time on Mir with her former role in Mission Control. 'I was a CapCom [capsule communicator – the astronaut in Mission Control who talks to the crew] many years ago for the first

Figure 6.6. The flexibility of a crew-member's schedule onboard a space station during a long-duration mission will be based upon experiences on Mir, and most probably as the situation demands on orbit. NASA has planned crew-members' schedules on short Shuttle missions down to the last detail, but on the International Space Station this will not be possible. Here, Dr Norm Thagard carries out experiments in fluid physics onboard the Space Shuttle in 1992. On Mir he found that due to a series of circumstances he had insufficient work.

German Spacelab mission ... I can remember watching some video of the people working on the Spacelab, and it was interesting, because every time I would make a call [up to the crew], the crew person would stop working and listen to the ground and reply. And I thought to myself, 'My goodness. This has a huge impact.' We ought to think about when we're calling up, because there's nothing more frustrating than trying to set up an experiment and then having to stop in the midst of screwing your screws together, or your cables together, and listen to what the ground has to say. They may be talking about something totally irrelevant to what you're doing. It deflects your focus.' Essentially, Dr Lucid recommends that the crew on orbit should generally initiate communications with the ground, based on what is required for their work.

Working on scientific experiments will be the purpose of having astronauts on the International Space Station. The lessons learned from the astronauts who worked on Mir have provided important direction for how the research work should be organised on the ISS. This includes an open and frank working relationship with both the NASA and scientific ground personnel, to ensure maximum efficiency and productivity of the astronaut's time.

6.3 THE STRESS OF LIVING IN ISOLATED ENVIRONMENTS

The psychological stresses that astronauts and cosmonauts have experienced from living in space can stem from pressures, internally and externally, to meet expectations and fulfil the objectives of the mission, under very difficult (and sometimes impossible) circumstances. Stress also arises from simply being in a small group, isolated from the normal social and intellectual contacts of everyday life. The effects of both sources of stress can easily become intertwined.

Regarding performance-related stresses, the 1998 NASA Phase 1 report recognises that 'more time should be planned for crew changeover, orientation, and experiment setup at the beginning of long-duration missions.' During the last Skylab mission, crew-members complained that not enough time had been allocated in the timeline for the setup of experiments. The consequence was that they fell behind in their work, creating frustration, conflicts with Mission Control, and psychologial stresses that could have been avoided had more time been allotted and had there been more flexibility in the schedule.

NASA learned onboard Mir that 'stowage and inventory control are also crucial areas for which adequate on orbit time must be allotted.' Nothing can be more annoying than wasting time looking for items that are needed but cannot be found. Mir astronaut Dr Andy Thomas emphasised the same point in an interview: 'There's a logistics issue that goes with being in zero gravity. If things are all stowed away in inconvenient places, you can end up spending a huge amount of time looking for things. And it's amazing how easily you lose things. You don't want to have a situation where you have to spend inordinate amounts of time looking for some stuff so you can do a 15-minute experiment. The way to overcome that is well-organised storage of all the tools that you're going to use, and the equipment that you're going to use.' NASA has recognised these problems, and is planning to organise the logistics and stowage on the ISS to facilitate a smooth working environment.

The stresses that result as a consequence of being in a confined and isolated environment – even if the work is progressing as well as can be expected – were in evidence during the astronaut increments on Mir, were observed twenty years earlier during the Skylab programme, and have been documented throughout two decades of Russian long-duration spaceflight. The symptoms most recognisable to space managers on Earth typically include a growing emotional distance, if not hostility, toward Mission Control, and a testiness that a crew-member would not normally exhibit in a comparable situation on Earth.

However, NASA seems to have been caught flat-footed when the Shuttle–Mir programme began. In her interview, Dr Lucid remarked: 'I think it's real important to include Skylab [in any discussion of the problems of long-duration spaceflight]. Sometimes we, especially here at NASA, have sort of forgotten that that occurred, it seems to me. I actually looked through the literature because I thought that it was very important that we not forget the experience of Skylab, and the only book that was ever written that I could find was *A House in Space*, aside from the NASA publications. I think it's a real shame that... it sort of dropped out of everybody's vision.' Henry F.S. Cooper's book, to which Dr Lucid referred, dealt extensively with the diffulties experienced by the last Skylab crew due to the pressures of being in space for nearly three months.

Remarking on the lessons to be learned from the experiences of the astronauts onboard Skylab, Dr Lucid commented: 'The very same things that all the Skylab guys said, all the Shuttle–Mir astronauts said, and we said, 'didn't you all ever listen to any of the people from Skylab?' You can go back and look at their debriefs. It was like living the same thing all over again.'

There are three principal methods through which psychological and social problems of small groups of space explorers can be studied and understood, and countermeasures developed for the International Space Station. One method is to study the previous experiences of crews in orbit, both the Americans on Skylab and Mir, and the Russians on their entire series of space stations over a period of twenty years. A second method is to examine the responses in situations that are similar to those faced by astronauts and cosmonauts, or analogues to spaceflight, such as scientific teams that 'winter over' in Antarctica. A third possibility is to study the adjustment to confinement and isolation of volunteers who are chosen for their similarity to real spaceflyers, and spend amounts of time comparable to space missions in isolation chambers that closely resemble the conditions that the actual mission will provide. All three of these approaches are currently being employed to develop methods of increasing the productivity, efficiency, safety and fun of the expedition crews aboard the ISS.

Having had the most experience with long-duration crews on orbit, the Russians have, by far, carried out the most intensive study of their reactions to the isolation of space. A Russian team of psychologists and behavioural scientists, the Psychological Support Group, is responsible for the emotional well-being of cosmonaut crews. After having examined the emotional and performance changes in cosmonauts over long-duration space station missions, they have found that during the first 2–6 weeks of a mission, the crew-members are busy adjusting to microgravity and learning how to perform their functions in this new environment. They are becoming assimilated to their new surroundings.

After the initial adjustment, the scientists observed that psychological problems, such as asthenia, arise, characterised by hypersensitivity, emotional irritability and instability, fatigue, sleep problems, poor appetite, and a decrease in the level of activity. Researchers studying Antarctic and submarine missions have described this second psychological phase of isolation and confinement as boredom and depression. In a paper presented at the congress of the International Astronautical Federation in October 1999, Dr A. Gregoriev and his colleagues, at the Institute for Biomedical Problems, summarise the characteristic of this period in space as one of 'an acute feeling of unsatisfied demands... the factor of novelty has lost its strength... and emotions are now mostly influenced by the factors of a drab environment and monotony, and forced company.' Depending upon the length of the mission, this phase can last for weeks or months.

The final phase that is observed is an hypomanic mood and aggressive behaviour, as return to civilisation draws nearer and the anticipation of the end of the mission

Figure 6.7. NASA learned from the Russians during the Shuttle–Mir programme that crew training for science experiments and other functions on the International Space Station will require that expedition members acquire general skills and a knowledge of specifics while on orbit. A crew-member will also have to rely more on his own ingenuity. Here, Cosmonaut Sergei Avdeyev, the flight engineer during the Mir 20 mission, enters data into a log in his quarters onboard Mir.

grows. Dr Gregoriev observes that this period can begin between two and four weeks before landing.

Dr Nick Kanas, working with a team of behavioural scientists from the Institute for Biomedical Problems in Moscow, has described as 'displacement' the negative interactions the crew begins to have with people on the ground, as a substitute for the expression of the hostility, anger or frustration they may be feeling toward their crewmates, but cannot directly express. It has often been observed that after an initial period of 'getting to know' one's crewmates, a colleague's small habits and idiosyncracies that were thought to be interesting or entertaining can simply become annoying.

In a 1998 article, Peter Freiberg cites the research that psychologist Joanna Wood – a senior scientist with Krug Life Sciences – has carried out, studying Australian-sponsored Antarctic missions for NASA since 1993. Dr Wood points out that the biggest problems during such long, isolated missions are interpersonal. 'After a few months you get tired of looking at the same faces. People frequently have behaviour that might be endearing in the larger society, but when you're living with it day after day, it's an annoyance.' She reports that a common complaint from Antarctic researchers is that people 'feel abandoned' by their home organisations. In space, during both the Russian and the US programmes, this has at times led to a deteriorating relationship between the ground and the crew.

As reported by Dr Kanas and his colleagues in a 1999 paper, during some space missions this friction and psychological distance between the crew and Mission Control has disrupted mission planning, and has led to problems in effective communication between the crew and mission managers.

A NASA-funded study that Dr Kanas and Russian and American colleagues conducted during five of the Shuttle–Mir increments, for the first time included questionnaires not only to help the determine the psychological state of the crew, but also the people on the ground. The study revealed that personnel working in Mission Control to support the astronauts were under considerable stress themselves, and lacked a sense of well-being. While this unhappiness was lower than in other work-groups in other studies, the authors recommend that 'more attention should be given to the needs of the Mission Control personnel.' Communication is a two-way street, and the state of mind of the person that the crew-member is talking to on the ground definitely plays a role in the emotional health of the astronauts.

This has also been stressed by Dr Al Holland, chief of psychology in NASA's Medical Operations Branch. As Peter Freiberg reports in a 1998 article, Dr Holland believes that actions taken by ground control and people in management can affect the emotional well-being of an astronaut, as much, or more than, the individual's psychological make-up. He says that problems can arise, for example, when the ground control personnel send up confusing or incorrect information to the crew. 'If these procedures are not clear or are erroneous,' he told Freiberg, 'they're really frustrating to the crew-member, who is already dealing with adapting to the environment... then you get tension between ground and crew.'

6.4 THE COMPLICATION OF CULTURE

During the Shuttle–Mir programme, and many of the Mir flights before and after, there was a factor to psychological dynamics among the participants, in addition to the generalised problems in all small, isolated groups. The study that Dr Kanas and his colleagues conducted during five of the Shuttle–Mir increments strongly reflects an added factor influencing the emotional balance of the participants in those particular missions: cultural differences. It should not be surprising that the researchers found, from questionnaires completed by five astronauts and eight cosmonauts, that the Americans were less satisfied with their interpersonal environment than were the Russians. The Americans were living in a Russian space vehicle, where only Russian was spoken, and where operational control of the mission was largely in the hands of their Russian counterparts. The astronauts had to function under a different operational philosophy and management style from that to which they were accustomed in the Space Shuttle programme.

In the study, the American astronauts scored lower than their Russian crew-mates in perceiving leader support and overall direction from their managers on the ground. The Russians viewed their environment as being more task oriented, more physically comfortable, and more encouraging of self-discovery than the astronauts. This, again, is hardly surprising, since language and cultural differences caused the

Americans to feel that their work was not well-defined, and, in later flights when substantial repair and maintenance work was required on Mir, some of the astronauts were unsure as to what their role should be. In an April 1999 article for the *Journal of the Society for Human Performance in Extreme Environments*, a team of scientists from Moscow's Institute for Biomedical Problems, led by Dr O.P. Kozerenko, detail some of the cross-cultural issues with which the Russians have had to deal, not only when there were American astronauts on Mir, but during the dozens of visits by cosmonauts from different nations before, during, and after the Shuttle–Mir programme. All of these issues have a bearing on future interpersonal relationships on the International Space Station, which will involve interactions between space explorers with at least half a dozen different primary languages and cultures.

The researchers stress from the outset that it is their view that 'several objective reasons (mainly, differences in the Russian and American approaches to space exploration) formed the basis for the development of the majority of the... phenomena' of problems of group interaction. The Russian approach is based on their experience of several-month missions with two or three cosmonauts, and the American approach is based on flights of between six and eight astronauts lasting, at most, a couple of weeks. This has resulted, they state, in 'the cultivation of certain different types of individual and group behaviour of crews in space'. As a result, there are different discernible types of behaviour in space, 'and the kinds of psychological problems which emerged.' The Americans, it has often been observed, are generally more individually career- and achievement-oriented, while the Russian crew-members are attuned more to the group mission and instructions from Mission Control.

At the beginning of the Shuttle–Mir programme, the American astronauts were on board as 'guests', with a specific complement of scientific experiments to carry out. Dr Kozerenko *et al.* point out that since the Russian cosmonauts had to spend the majority of their time on station repair, the 'guest' a cosmonauts 'had to make the difficult choice between giving priority to scientific experiments in their flight programmes while ignoring the difficulties of the station 'hosts', or to pitch in and help repair the station with the Russian crew, to the detriment of the American flight programme.'

Each astronaut had to make that choice, especially after the February 1997 fire on Mir, and those who took an active part in the repair operations together with the cosmonauts, became psychologically 'closer', they observe. It can easily be imagined that a similar situation could arise on the International Space Station, which will be largely staffed with Americans and Russians, but which will include other astronauts, sent aloft to conduct experiments for their respective space agencies.

In Dr Lucid's case, even before there were serious repair issues on Mir, as Dr Al Holland has described, 'if it came down to choosing between the two [social or scientific activities] she would let the work wait and concentrate on the social side of things.' In that case, Dr Kozerenko points out, 'Shannon Lucid's flight vividly demonstrated that language barriers, cultural and gender differences, variation in perspectives, and so on, could be overcome if cosmonauts and astronauts gave

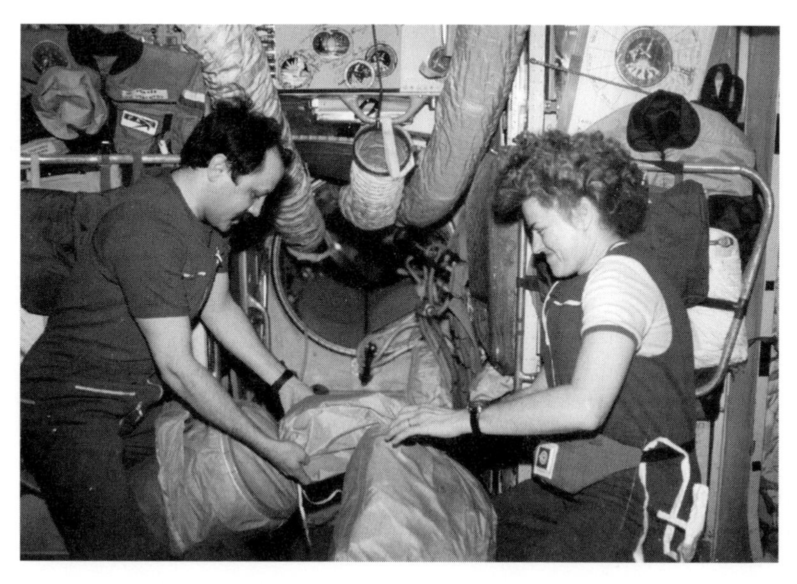

Figure 6.8. Dr Andy Thomas and other Mir residents have pointed out that habitability onboard a space station requires that essential equipment is readily accessible. Stowage of supplies became a problem on Mir as additional Progress supply ships and the Space Shuttle made deliveries to the station. Soon after arriving at Mir, Dr Shannon Lucid helped flight engineer Yuri Usachov move supplies to their proper places from the core module. Note, in the upper centre, the mission patches which have been signed by each of the visiting Shuttle crews.

attention to the social side of things, showed empathy, and played a proactive role in the smooth adaptation of station crews.'

A similar choice faced astronaut Dr Michael Foale, who was present on Mir when the Progress supply ship collided with the station, puncturing a hole in the Spektr module and leading to its depressurisation. The three-man crew immediately understood that they were in a life-threatening situation: if the leaking module were not isolated, all of the atmosphere in the station would be lost. According to regulations, and as a safety precaution, when the Russian flight engineer Alexander Lazutkin realised that the Progress was going to hit Mir, Dr Foale was instructed to move quickly to the Soyuz spacecraft – the return vehicle that is always stationed at Mir.

Dr Foale reports that the crew could recognise from the rate of the change of pressure in their ears that the rate of depressurisation in the Mir was not very high, and that there would be time to try to isolate the leak rather than abandon the station. He saw that his two Russian crew-mates 'were not coming quickly to the [Soyuz escape] spacecraft.' The cosmonauts knew where the leak was, he reported, and Lazutkin was 'rapidly trying to dismantle the cables [to the Spektr module], so I then [left the Soyuz] and helped him.' For the rest of his increment on Mir, Dr Foale became an integral part of a team working to restore the functioning of the station. He also actively intervened within the group to try to restore some confidence to the commander, who had suffered the humiliation and self-blame of having been in charge during the accident.

Operational culture and training issues played a significant role in the psychological impact on the crew in dealing with various of the crises onboard Mir. Dr Kozerenko states that since the short flights on the Space Shuttle are 'technically reliable', the astronauts 'concentrate on the strict and timely execution of planned tasks, and are surprised that cosmonauts, in some cases, change their daily schedule to give more priority to technical operations. This difference in perspectives can be the basis of conflict during distribution of work resources and competing tasks.' While the American view is more attractive to scientists, 'during accidents, it could be argued that the Russian flexibility and readiness for any change works the best.'

The Russian team warns that the isolation felt by space crew-members who are 'different' from the majority (such as when one American was with two Russian cosmonauts; Dr Thagard referred to himself as feeling like the Lone Ranger), can also be an issue on the International Space Station. The ISS will have modules built and staffed by different nations. 'National, cultural, religious, professional and other differences could be the basis for the formation of subgroups, having their own interests and regarding others as different.' They point out that 'the experience of the joint Russian–American polar expedition testifies that conflicts of this type are more difficult to resolve than simple conflict among individuals. The Russian leader supported the priority of safety in activity planning, while the American leader insisted on rapid gains in goals.'

Dr Kozerenko and his colleagues have compiled a list of suggested methods of overcoming the cross-cultural problems that could arise on the ISS. One of these is to train all flight crew-members in flight operations together. Training in extreme conditions, such as survival training, has been mentioned by American astronauts as an effective way to increase confidence among crew-members. There could also be a reduction in the influence of cultural, national, personal and other factors on performance, if sensitivity training were included, by experienced social and behavioural experts.

During the Shuttle–Mir programme there was another, more objective factor that led to some amount of friction between the Russian crew and the Americans on board. Because the replacement of cosmonauts on Mir did not generally coincide with the replacement of the astronauts, a number of the astronauts served with more than one Mir crew. It was not always possible to hold training sessions on the ground with all of the crews that the astronaut would accompany on Mir. The Phase 1 Joint Report states: 'The result was that in some flights the crew commander, without knowing the actual proficiency level of the astronaut, did not always trust the astronaut to perform individual flight engineer-2 operations, even when the latter was adequately trained to do so.'

Dr Kozerenko *et al.* also importantly recommend that 'the opinion of crew-members should be solicited and taken into account regarding everyone's access to logistics, information and communication tools, workload distribution, and so on, so that potential sources of quarrels can be identified and offset.'

At the same time that crews are beginning to be trained for the three-month expeditions on the ISS, studies on the ground are continuing in the United States and Russia, with participation from the European Space Agency, Canada and Japan, to analyse the interaction between small groups of people in simulated space station environments, in order to provide more 'lessons learned'. It affords the space agencies the possibility of testing the efficacy of various methods for mitigating the traditionally observed psychological effects of long-term isolation, in order to provide the optimal conditions for future ISS crews.

The most recent study is being conducted by the Institute for Biomedical Problems in Russia. In July 1998 the Institute called for volunteers to participate in a unique 240-day experiment of flight simulation aboard the ISS. It is the longest test of its kind ever conducted, and takes place in an isolation chamber designed to be a high-fidelity simulation of the Russian module for the ISS.

The eight-month simulated isolation test began on 2 July 1999, with a crew of four subjects including two Russian cosmonauts. During the simulated mission there will be 89 scientific experiments undertaken, designed by all of the ISS partners. The experiment is multi-fold in purpose. The subjects are testing the durability of hardware that will be launched into space, and the researchers will be able to evaluate how their equipment performs. The subjects' interactions with each other and with 'Mission Control' will be observed, as well as the activities they choose to combat the typical response to isolation.

Figure 6.9. A sure sign that crew-members on a mission are suffering the effects of isolation is a deterioration in their communications with Mission Control. Dr Nick Kanas and other researchers have recommended that there be attention paid to the stress on ground personnel as well as the space crews, and that improved training among all of those critical to a mission be implemented. This is Mission Control in Korolev, outside Moscow.

M. Ephimia Morphew, President of the Society for Human Performance in Extreme Environments, reported in a personal communication, on 14 September 1999, on the progress of the test subjects after their first two months in the hermetic isolation chamber. She reports that the crew completed 98% of the planned scientific protocol, with the incomplete tasks being due to problems with hardware and software. The crew also expressed an interest in growing plants in their chamber's greenhouse. It was found that some scientific equipment did not meet the rigours of long-term use in space, and improvements will be designed. The researchers report that during the first two weeks of isolation, each crew-member established a relationship with Mission Control. During this period, Morphew reports, the crew 'felt time deficit and complained about fatigue. But as they adapted to the experimental conditions, there are no more complaints about work schedule.' They are managing to complete the daily work schedule on time.

The Russian scientists report that at the two-month mark, the crew's interaction with Mission Control is good, and that the psychological support group has arranged that special occasions, such as a birthday, be celebrated with family and friends. Each crew-member has access to electronic mail and the Internet, and 20 minutes each week for telephone communication.

In addition to the analytical studies, referred to above, that have been undertaken to better understand the problems that arise in group interactions in isolated environments, the astronauts themselves have been quite outspoken about the problems they encountered on Mir, and their recommendations for changes to be made on entering the era of the International Space Station.

6.5 ANALOGUES TO SPACEFLIGHT

In his exhaustive and fascinating study of the lessons to be learned from polar expeditions for space exploration, Jack Stuster has highlighted some additional factors that effect the physical and psychological well being of explorers living in isolated environments. These lessons from polar exploration, and practical suggestions, can and should be considered for the International Space Station.

During the long periods of total darkness at the poles, in the mid-phase of a polar expedition, sleep disturbances have been noted, as they have also been observed among long-duration space crews. Stuster suggests adherence to regular schedules of sleep and activity. This was a problem on the last Skylab mission and during crisis periods on Mir, when crew-members forewent sleep to try to regain time that was lost during the day.

Stuster also recommends that tasks should not be scheduled for one hour before the sleep period, so that crew-members can relax before going to sleep. Sleep quarters should be well-insulated from noise and light, and sleep chambers can be outfitted with communication devices to alert the crew to any danger. Considering the difficulty described by astronauts on Skylab of responding to a call from Houston, in the dark, such devices should be easily locatable.

Many debates have been carried out as to the efficiency of exercising in microgravity in mitigating the effects that the environment has on human physiology, particularly on the deterioration of muscle and bone (discussed below). While how much exercise and which kinds are appropriate may be open to debate, there is little question that for a long time to come, until perhaps other countermeasures are developed, astronauts and cosmonauts will be spending significant amounts of time exercising in space.

Working out on a treadmill or bicycle, for several hours each day, has understandably been regarded, by astronauts and cosmonauts alike, as an unpleasant activity. Stuster recommends that if people are required to exercise, it should in some way be enjoyable. While polar explorers had the option of skiing, and space explorers will have more limited options, he offers some novel methods. One possibility is to connect the exercise device to an electrical generator that contributes to the power requirements of the spacecraft, to provide the crew-members with some tangible results from their exercise. Crew-members can also maintain personal performance records so that they have a sense that progress is being made over time.

There can also be a recreational exercise which is not simply viewed as a chore. Skylab astronauts recommended, for example, that the exercise device be placed in front of a window, so that the crew-member could enjoy the outside view, which they would otherwise see only during their free time. There could also be a television with variable programming placed near a bicycle, or other device, and crew-members could listen to music while exercising. There should be an exercise area which is a dedicated 'mini-gym', Stuster suggests, and there should be more than one type of exercise device used, on a rotational basis, to offer variety. He puts forward the novel idea that consideration be given to rotating the devices for exercise, to correspond with seasonal changes on Earth.

The style and manner of the leader of any expedition has been well studied, in terms of the impact that it can have on the success of the entire mission. Many studies have shown that while crew-members work best when there is someone who is always 'in charge', a team approach has often meant the difference between harmony and mutiny. Stuster includes the following quote from a personal communication from Thor Heyerdahl, who led sea voyages during his Kon-Tiki (1950) and Ra (1971) expeditions: 'A successul expedition is the result of thoughtful planning and harmonious teamwork. Each person must feel that he or she is a necessary part of the team, and that a successful outcome depends on them, but neither more nor less than on everyone else. There must be a feeling of friendship and equality where, nevertheless, everyone realises why the leader is a leader among equals; because the leader is the one who knows best what it is all about. Thus I have always made it a point to sit down with the men, listening to everyone's opinions before important new decisions are made. Only when there is a lack of time, and a decision must be made immediately, have I given orders like a military officer.'

Stuster further recommends that the leader of a long-duration space mission must be neutral regarding controversial issues, and possess 'exceptional interpersonal skills'. While the leader must be ready to take personal responsibility for the mission, he must also exhibit concern for the well-being of the crew, and directly interact with each member of a small group.

Figure 6.10. Dr Michael Foale, and the other astronauts who lived on Mir, had to make decisions as to how they would interact with their Russian crew-members. By maintaining a sense of emotional equilibrium, and concern for the psychological stress on the cosmonauts after the collision with the Progress ship, Dr Foale was able to contribute to the completion of his mission to Mir, and provide support for the crew. After four months onboard Mir, Dr Foale appears ready to come home on the Shuttle.

During a long-duration spaceflight or an isolated increment in a polar station, food can take on a social quality not always evident back home. In addition to providing nutrition, meals are a social time for a small group, and an opportunity to spend relaxed time away from the toil of the day. Stuster suggests that in recognition of the social value of preparing special meals for special occasions, which reduces the routine nature of the mission, the design of the food system should accommodate 'more elaborate meal preparation when schedules and workloads permit'. Special dinners should be encouraged, he suggests, and galleys and dining areas must comfortably accommodate all members of the crew. He also warns that mission managers should 'expect complaints about the food regardless of how much attention is devoted to ensuring that it is the best possible under the circumstances.'

Related to the question of food in space is the question of the overall aesthetics of the habitat in which the expedition members live. Here the factor of individual variability and overall psychological adjustment to isolation are paramount. While some astronauts complained that the battleship grey colour of the walls of Skylab were boring, if not depressing, others found it soothing and relaxing. Logic would dictate that while crew-members should be encouraged to decorate their personal quarters, the personalisation of decor in commonly used areas should be discouraged.

It has been noted in studies of crew on submarines and at polar stations that the preferred artwork to adorn the habitat is of wide vistas that open the viewpoint of the crew-members beyond their small confines. Stuster recommends that scenes that are particularly pleasing should be displayed, with a selection of images that can be rotated, perhaps according to the change of seasons. He recommends that there be a variation of visual stimuli through colour, and the effect of colour has been recognised by the Russians as one way of intervening in the state of mind of crew-members.

In 1999 *Space News* reported that Mikhail Kuzmin, chief of optics at the Institute for Biomedical Problems in Moscow, has devised a set of coloured lenses to be tested by cosmonauts in space, to help alter their moods. Green lenses will be put on to help them relax, and orange–red lenses will be used for energising work. Preliminary tests with the coloured lenses, on volunteers at the Institute's simulators, produced higher levels of concentration when wearing the organge–red lenses, and participants made fewer mistakes. Researchers at the Institute note that colour therapy has been used in many cultures for hundreds of years, although they are not optimistic that the 'foreigners' (the non-Russians) on the ISS will be willing to participate in this programme.

Studying the 'bold endeavours' of men throughout the history of exploration has provided insights into the problems faced by small groups of explorers facing extreme environments. We also have the observations and suggestions of the Shuttle–Mir astronauts themselves, who have already had a taste of the some of the issues that will be faced on the International Space Station.

6.6 BEING THERE

It is well known that in the rush to ready the first astronaut to live on Mir, Dr Norm Thagard, scant attention was paid to the possibility that he would become psychologically effected by the isolation from Earth, family, friends and colleagues, and by the added social distance of being American and English-speaking, unlike his two Russian crew-mates.

In an interview with Dr Thagard, published by the American Psychological Society and reprinted in *Human Performance in Extreme Environments* (HPEE) in 1997, the first Mir astronaut discussed the psychological difficulties experienced on his mission. The communications between Mir and Moscow Mission Control were only in Russian, as were Dr Thagard's interactions with his crew-mates. And while he considered himself proficient in his new language, he commented that 'from a cultural standpoint, it's an isolated environment.' Receiving English-language newscasts from Earth had not been built into the daily routine on Mir.

In further comments to *HPEE*, Dr Thagard explained that he received 'no news for the first five weeks of the mission, although my Russian crew-mates were getting regular [Russian] news broadcasts almost daily.' Actually, he continued, 'I had not even thought about the lack of news until, one month into the mission, an interviewer asked me if I was getting the news up there. I replied with a simple 'no', and wrote in my log that this represented cultural isolation that should be avoided on long flights with multinational crews.'

Upon return, Dr Thagard recommended that there should be a 'formalised provisions for [such] communication in whatever language is needed,' despite the fact that the official language of the ISS will be English. Accommodation should be made, he suggested, for those whose first language is not English. This issue has also been raised with regard to safety issues that will arise on the ISS. The NASA–RSA Joint Report on the Phase 1 programme points out that due to the shortness of training time, it was not possible for the astronauts to become as proficient in Russian as everyone would have preferred. Although longer training will be the goal for the ISS, the recommendation is that for some 'time-critical and safety-critical procedures to be absolutely clear and easily understood in an emergency, some minimal amount of multi-language procedures may be required.'

Dr Thagard expressed his frustration at 'not having enough to do' for a period on Mir when the experiments he had been trained to execute had been postponed due to the late arrival of the Spektr module, equipment malfunctioned, and other factors were taken into account. In general, while the fear has been that station crews will be overburdened with work, the opposite could easily be the case, under similar unavoidable circumstances. Dr Thagard stated that to him, 'the single most important psychological factor on a long-duration flight is to be meaningfully busy. And, if you are, a lot of the other things sort of take care of themselves.' His recommendation that NASA must 'pay attention to [psychological] factors', did not go unheeded by the space agency.

After Dr Thagard returned from Mir, NASA Administrator Dan Goldin made a public apology to the astronaut. After the Shuttle landed bringing Dr Thagard home, Mr Goldin said: 'We put all of our focus on the physical well-being of the astronauts and the success of the mission. We neglected the psychological well-being, and Dr Thagard made it clear to us.'

In addition to providing support for crew-members once they are in space, many of the Shuttle–Mir astronauts insist that problems can be avoided, or at least mitigated, by including personality and psychological criteria in the selection of astronauts for expeditions on the ISS. Dr Thagard suggests that NASA have on its agenda 'attention to crew interaction and compatibility, and you have to actually look at the psychological status of the individual crew-members.' While this is now carried out to some degree, Dr Thagard believes that NASA must increase its efforts to select crews carefully.

This evaluation was strongly echoed by Dr Lucid, who stressed this in her interview, drawing from her own experience: 'I think the most important thing is the composition of the crew. First, the person has to be suited for long-duration space flight, and second, the mix of the crew has to be a good mix. You can have four or five people, and each one individually would be absolutely a great person to fly in space for long periods of time, but if you put the wrong mixture together, then it could be a very miserable experience... I was very, very fortunate, because there could not have been two better people to fly with than Yuri [Onufriyenko] and Yuri [Usachov]... We worked well together. We complemented each other personality-wise.' When Dr Lucid was asked for recommendations for long-duration spaceflight, she said: 'The most important thing is the mix of the crew.' Is NASA taking that recommendation into account? 'They're paying a little bit of lip service to that,' she replied, 'but I don't know exactly. The reality of life is stronger than what might be a theoretically nice thing to do. In reality, you'll never be able to pick perfect people, or perfect crews. So what you also need to do is have a plan.' After a crew is chosen, she suggested, the question should be 'What can we do to help them have a great time? I think there are ways to help crews so they can facilitate working with each other and enjoy the time together.'

Although Dr Lucid does not think that considerations such as the human

Figure 6.11. The Mir astronauts, and both Russian and American specialists who have worked on combating cultural differences among international crews, all note that training crews together extensively, especially in extreme environments, helps to improve cohesion among the crew-members, and tends to eliminate conflicts of culture. Here, the second crew that will live onboard the International Space Station – Susan Helms, commander Yuri Usachov, and Jim Voss – undergoes winter survival training near Star City in Russia in March 1998. The training prepares the crew for return from the ISS during an emergency in a Soyuz spacecraft, and for landing where they might not immediately be reached.

interactions of the crew will determine how individual groups of astronauts will be chosen for a specific crew, she counsels that 'it doesn't take a PhD in psychology to figure this all out. Looking around the [astronaut] office, just about anybody can say, whenever a Shuttle assignment is made, for instance. 'Oh, what a great crew that will be', because you know the people and you know they are going to interact. Or, sometimes, when a Shuttle crew is announced, you think, 'Wow! Who dreamt up that mix of people!' '

The empirical proof of Dr Lucid's analysis is the fact that three years after she set the record of being the American with the longest time in space, according to comments to the Associated Press on 21 September 1999, she is anxious to return to space. It is well recognised that it was the force of her good-humoured and joyful personality that was critical in making her mission to Mir a resounding success. Certainly not many people would have handled, with such grace, being told that their stay on Mir would be extended by more than a month because the Space Shuttle was temporarily grounded. It would even have been understandable if an astronaut in that situation complained about it to Mission Control, which was not Dr Lucid's way of handling the situation. Dr Al Holland, chief of psychology at NASA's Psychological Services Group remarked, regarding Dr Lucid's longer stay aboard Mir: 'I think if it had been another crew-member, we would have heard complaints.'

In an interview in December 1998, Dr Andy Thomas, the last astronaut to live on Mir, also commented on the qualities that he thought were important for a long-duration crew-member: flexibility and independence. 'I think people need to be flexible', he advised, 'to have an adaptability to new situations. I think you also need to be able to creatively find recreation for yourself, sort of within yourself, and not be too dependent on externals, particularly other people... I think, for example, that if someone has a social life that's very outgoing, and they have a great dependence on social contacts with lots of different people, obviously they're going to have a difficult time in the confinement of a space laboratory.'

The criteria used for the selection of astronauts are under active discussion in the space community. Reviewing the successful versus miserable expeditions of explorers in the past, Jack Stuster found that social compatibility was deemed more important than technical competence in choosing participants by the leaders of the successful missions. In selecting their men, they were looking for those who are compatible with others under difficult conditions. And among those who completed winter-over missions in Antarctic stations, it was found that they ranked emotional stability and compatibility as the most important personality traits for successful performance.

Personality has not been the primary criterion used for astronaut selection in the American space programme. Technical competence has always been the overwhelming qualification for spaceflight, with astronaut candidates only 'selected out' for pathological psychological conditions; they are not 'selected in' on the basis of the identifying positive personality traits that will be important for mission success. Stuster suggested that high-fidelity simulation studies of astronaut crews would reveal at least some of the problems that may be likely to develop during a mission.

In addition he recommends that while technically competent candidates could be

Figure 6.12. Living in space is not the first venue where people will be subjected to isolation in an extreme environment. Much can be learned from studying the analogues to spaceflight, such as the expeditions to the North Pole and Antarctica. Remote camps, such as this camp in Antarctica, can be reached only by aircraft or helicopter, and only when the weather permits. The tents are similar to those used by Robert F. Scott during his attempt to be the first explorer to reach the South Pole. (Courtesy Bert Rowell/ National Science Foundation.)

identified, individuals should be selected who exhibit the appropriate personality traits for long-duration missions, such as social compatibility, emotional control, patience, tolerance, self-confidence (without being arrogant), flexibility and practicality.

In her comments to AP, Dr Lucid also pointed out that it is very important to long-duration crews for the ISS to ensure that it is the appropriate time for them to make such a commitment. 'It would have been very, very difficult [for me] if I'd had small children at home,' she remarked. Since her youngest child was in college at the time of her Mir mission, she knew they 'certainly didn't think they needed a full-time mother.'

Mir astronaut Dr Michael Foale made a similar point in an interview. When, in May 1998, I asked him about his plans for the future, he replied: 'My plans are to keep doing what I've been doing, which is enjoy my family, bring up by children responsibly, and continue to get people into space – and, if possible, myself. I would like very much to go back into space, and do space-walks, which I love doing. I got to do one on Mir, as well as once on the Shuttle... And then, after that, in maybe two years, I would like to go to the International Space Station, and do another long-duration flight. And I choose that timeframe, because that's when my son is going to be six or seven. He'll be well on his way to having his own character, without any regard to my influence. I feel [then] I can kind of excuse myself, even if my children don't excuse me, for going away for six months.' Dr Foale is to participate in the third mission to service the Hubble Space Telescope from the Space Shuttle.

No matter who is selected for long-duration station expeditions, ground support will still be required. Although the space agency was not prepared to provide adequate psychological and social support to the first astronaut on Mir, during the eight months between Dr Thagard's return from Mir and Dr Lucid's launch to it, NASA instituted a ground support system which improved communications for the Mir astronauts that followed. As NASA states in its 1998 Phase 1 report, the agency recognises now that 'it is essential to address psychological factors early to maintain crew morale and efficiency throughout long-duration stays.' They admit that critical psychosocial support 'turned out to be much more important than initially expected and will be emphasised for the ISS.'

Figure 6.13. Christmas on Mir. The Mir astronauts have stressed – and the Russians have long known – that every day must not be like every other day when spending months on a space station, away from family, colleagues and friends. The Progress supply ship periodically delivered letters, special foods and treats for the Mir crews. Birthdays and holidays are also important distractions from the monotony of the surroundings. Here a space Santa Claus graces Mir for a holiday celebration being enjoyed by the crew. Twenty years before the Shuttle–Mir programme, the Skylab astronauts recognised the same need to celebrate a special day; but they did not have the advantage of deliveries from Earth, and on 24 December 1973 they fashioned a Christmas tree out of food cans (inset).

6.7 MITIGATING THE EFFECTS OF ISOLATION

The Russians have long realised that a psychological support group on the ground is a necessary part of maintaining a cosmonaut's mental health on orbit. One of the responsibilities of the cosmonaut Psychological Support Group has been to help monitor and organise the leisure time and interactive activities of crews on long-duration missions. While the US communications have most often been limited to family and reporters during on-orbit news conferences, the Russians provide cosmonauts with a range of famous and accomplished public figures to talk with, including sports celebrities, poets and musicians. In this way the cosmonaut feels that he is being given special treatment and attention, and that what he is engaged in is of interest to a wider group of people apart from the space agency.

Russian mission planners have also routinely included personal letters, newspapers, books, favourite foods and surprise gifts for the comsonauts onboard unmanned Progress supply vehicles that serviced Mir. (Jack Stuster reports that cosmonauts have commented that the letters they enjoyed the most were those from school children.) Indeed, it was found in one simulation study that the delivery of such goodies to the subjects in isolation lifted their spirits, effectively combatting the usual effects of boredom, monotony and interpersonal tension.

Other leisure activities that the cosmonaut Psychological Support Group have prepared are a range of resources including audio and video tape libraries of music, movies, and lectures, assembled with the preferences of the individual crew-member

Figure 6.14. Dr Andy Thomas, the last astronaut to live onboard Mir, stresses that a crew-member must be able to make a 'home' for himself in whatever living area is available for his personal habitation. He has been a strong promoter of increasing international participation in space exploration, and is seen here, after his mission to Mir, with IAF president Dr Karl Doetsch, at the Congress of the International Astronautical Federation in Melbourne, in his home country of Australia, in September 1998. (Photograph by Marsha Freeman/*21st Century Science & Technology*.)

in mind. As Drs Lucid and Thomas also stressed, the space traveller should plan ahead of time what leisure activities and forms of communication he prefers, so that they can be prepared in advance of the mission.

The Russians also typically monitor the psychological state of the cosmonauts during a mission. Dr Kozerenko *et al.* report that the Institute for Biomedical Problems carries out a computer analysis of e-mail messages sent by crew-members, to determine the state of mind of the subject. Whether such an activity would be accepted by American or other national crew-members remains to be seen.

The NASA report on lessons learned from Mir states that the 'Phase 1 experience reinforced the need for a customised programme of psychological support for each ISS crew-member. Family members were included in pre-flight psychological preparation and training for later Phase 1 missions.' In the opinion of the Russian and American crew-members, joint training sessions under extreme conditions (survival training) would also contribute to developing working relationships. To try to avoid, or at least mitigate, the frequently observed strained communications that develop between a crew and the 'outsiders' on the ground, the report states that 'pre-flight team building between the crew-members and the ground personnel was also found to be important for a good working relationship later on.'

In space, and in other isolated environments, it has been found that boredom can be as depressing as too hectic a schedule, as Dr Thagard has reported. Lack of productive activity can result from equipment breakdowns, or by simply completing tasks in less time than allocated. And the monotony of talking to the same people, and having only a limited variety of diversions, can be very psychologically wearing. Dr John Uri, who was responsible for the science programmes during the Shuttle–Mir mission, reported in an interview that from his debriefings of the Mir astronauts, he concluded that 'boredom is probably the worst enemy up there, of any crew-member... It's not like you can go out to the store, or the mall, or go to a movie... The boredom is one factor that is worse than being too busy. We heard that from a lot of crews: 'During those time periods where I wan't busy enough, was not a good time.' Your mind needs to be focused. These [the astronauts] are all very high-achieving kind of folk, who need to have their minds occupied with, what is to them, useful work... We're very much aware of that,' Dr Uri continued, 'through the assembly phase, because we have a limited amount of science that we can do, because the construction and assembly take up the majority [of the time], and are the highest priority. But, supposing for some reason, all of a sudden there is some free time that becomes available. How are you going to keep those crews busy? For instance, what if a set of EVAs has to be deferred to the next mission, or the Space Shuttle that is coming to pick them up is delayed by a couple of weeks, or six weeks? How are we going to keep them busy when their mission is actually over, but they are still up there with nothing to do? That's a hard one to puzzle over.'

The question is, he said: 'What can you take up there that can sit there for months, and all of a sudden, be taken out of the closet and worked on? Can you fit it on the Shuttle, to take it up to orbit? Is there room to stow it somewhere, in addition to the regular stuff you need?' Dr Uri believes that 'it's a complex issue, but I think from the psychological standpoint it's a very important one. To have anyone up

Figure 6.15. Just as on Earth, there must be a way for a crew-member to 'get away' from the daily routine, and even the company of his companions. Dr Shannon Lucid amassed a collection of more than one hundred books during her stay onboard Mir, and here she is seen enjoying one of them in the Spektr module. The astronauts stress that productive and creative leisure activities are critical for psychological balance. (Courtesy NASA and the Russian Space Agency.)

there twiddling their thumbs doesn't do anybody any good. It's got to be useful time.'

The astronauts have had some suggestions in this regard. In comments to *AP* in September 1999, Dr Lucid reported that she counsels crews in training for long-duration expedition missions on the ISS to find a hobby to help occupy their time in space. During her Mir mission she spent her free time reading the 100 books that she had her daughters send up while she was there.

In his book, Jack Stuster explains that the English word 'recreation' is derived from the Latin for 'restore to health'. Dr Thomas has stressed that hobbies and activites for relaxation are critical for the mental health of the astronaut. 'The big challenge' on a long mission, he stated in an interview, 'is to find a way of psychologically removing yourself from an environment, when you can't physically remove yourself from it. And that's why creative recreation is so important. It lets you do that.' It also provides a 'personal reward from your activities, and that can break the monotony.'

Dr Thomas reports that when he returned from his Mir increment, 'one of my recommendations was that NASA provide certain standard forms of recreation, such as music, compact discs, movies, computer aids, and so on. But I think it's important that the individual have something that is personally rewarding to him, if such a thing exists, and it is transportable.' Dr Thomas therefore suggested that 'a personal recreation vehicle, whatever it is, needs to be accommodated by NASA. In my case it was providing pencil and paper,' since he had decided to draw, 'but it might be providing a musical instrument, or a particular collection of books. I think that could go a long way to making a very big difference in the quality of the time that the person has in orbit.'

During American space missions there has always been the pressure, on a short flight, to 'get as much work out of the crew as possible'. Certainly the astronauts want to complete as much research as they can, and NASA has always crammed their schedules with experiments. But Dr Thomas points out that on long-duration missions a more regularised work schedule has an important bearing on the necessary psychological relief from the day-to-day routine, just as it does on Earth.

'We actually had our time set up as a sort of work week [on Mir], much like a five-day work week. The weekends we still had work to do, we still had duties, of course, they didn't go away, but it was a reduced level of duties. So the weekend gave us a lot of time just to do things like take pictures, look out the window, or to do some recreation, to watch movies. In my case, to do some drawing, or write mail, or reading. And there was something kind of nice about having weekends up there. It broke the monotony of every day being like every day. And it has some sense of normalcy, or the kind of routine you're accustomed to [on Earth]. It lets you get psychologically recharged for the coming work week, much like it does here on the Earth.'

Dr Foale has reported that the regular Saturday movie night during his Mir increment not only provided a relaxing social activity for the entire crew, but that this was an especially important relief during the highly tense period following the collision of the Progress with Mir, when it was necessary to support the mission's commander.

In addition to recreational activities that the crew can share together, such as watching movies, Dr Thomas stresses that 'eating meals together is very important. Your schedules have to be matched to accomodate that, so people can all gather and share meals together, and have that sort of social interplay. That goes a long way to having a sort of balanced life up there.'

In his book, Jack Stuster provides some additional ideas for recreational opportunities, drawn from his studies of life in other isolated and extreme environments. He suggests that both personal and common facilities should be provided for listening to music. And while the technology exists today for each crew-member to listen to music, or watch television and video tapes in private, leaders of expeditions have tried to encourage as much group activity as possible.

A selection of card and board games, which have not, to the present time, been very popular in space, could be included on a long-duration mission. Formal courses of study and organised weekly lectures by crew-members themselves have been found to be enjoyable during polar missions, as well as experiments chosen by the crew-member to pursue for his own enjoyment. This integration of work with leisure has been noted in the special attention which space explorers have given to growing

Figure 6.16. One of the most important antidotes to psychological stress and feelings of isolation is regularly engaging in social activities, as a crew. Work must take second place, crew-members have suggested, to eating meals together, for example. In this photograph, Mir 18 crew-members Vladimir Dezhurov (left) and Gennady Strekalov enjoy a meal onboard Mir.

plants in space. It is also operative in making the necessary meals pleasurable experiences for the participants. There may even be the possibility of developing an expedition newpaper to inform and entertain the crew, Stuster points out.

All of the studies, however, show that the major avenue preferred by crew-members to combat feelings of isolation and home sicknessess, is communication with friends and family on the ground. In her interview, Dr Lucid stated: 'I think e-mail is the most personal way of communicating. That's the way I kept in contact with my family when I was in Russia while I was training, because there was no phone. I would spend hours trying to make a phone call back to the [United] States. Most of the time, it didn't work. I got e-mail messages from every member of my family every night. I sent them e-mail messages. They talked about what they did that day, so I was totally in sync with their daily lives.' Dr Lucid continued to reply on electronic mail on Mir.

She recommends that each individual person should be able to choose what form of communication they prefer to have with family and friends. 'I think the system should be flexible in adjusting for each individual crew-member. I also think it's very important that the crew-members realise ahead of time what is important to them, so the system can provide that.' Dr Thomas concurs. In his case he was not anxious to have telephone or video conversations with family or friends on Earth, but preferred electronic mail. 'I've heard it said,' Thomas related, 'that people who go off to West Point, or a military academy, go through the same thing. They like to get letters, rather than phone calls, because a phone call pulls you back to the environment that you have to pull yourself away from.' Dr Thomas feels that the expedition crew-member must try to function as productively as possible 'without trying to live in two worlds at once.'

In an environment where the number of square metres is very limited, astronauts have found that it is important to agree amongst themselves to create an individual 'psychological space' within the limited physical space available. The physical design of the space station must accomodate that need. Dr Thomas stated that 'the most important thing is that you do have some region in the spacecraft which you can basically call 'home', and you think of it as your place, where you have personal effects, your recreational aids. It's a place where you go. It might be where you sleep. In my case it was the place where I slept, and also the place where I worked... This gives you a sense of somewhere that you belong, that you can go to when you need to be alone, or when you wish to be alone. You know that the area will be respected, and that if you want privacy, you will have it when you need it. And that certainly happened on Mir. It makes a huge difference that if you leave something such as your personal effects there, and you go away to do work in another module, when you come back ten hours later, you know that it's still there. It gives you the sense of coming home. That makes a big difference.'

6.8 THE BODY AND THE MIND

NASA reports in its Phase 1 evaluation that Russian and American medical experts combined forces during the Shuttle–Mir programme to provide an in-flight system of health care that took advantage of the best of what each programme has developed in 40 years of taking care of crew-members in space. For the past twenty years, Russian medical professionals have faced many health issues during long-duration spaceflights.

Medical emergencies due to accidents, and a variety of illnesses, will undoubtedly have to be treated, to some degree, aboard the ISS. During the Shuttle–Mir programme, NASA developed a capability in Houston for astronauts undergoing training in Russia to have access to medical experts whom they knew back home. A telemedicine system was linked to NASA's medical facilities around the United States, which allowed astronauts to consult with experts in real time.

NASA has been working with an international group of physicians to enhance the capabilities available to crew-members in flight and to also stay abreast of new developments in health and medicine. In order to enhance their ability to interact with crew-members, NASA has also been investigating the use of the Internet as a tool in improving crew health-care. During the Shuttle–Mir programme, astronauts and cosmonauts were in contact with doctors on the ground through both a visual and audio communication system.

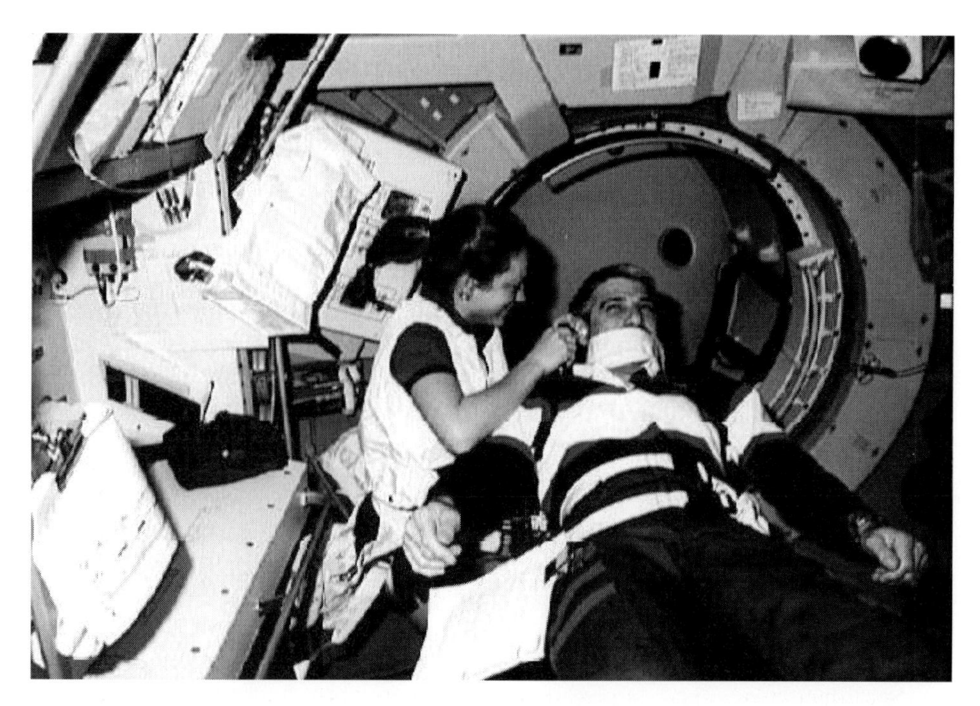

Figure 6.17. Crew health has been of primary concern from the time the first space travellers stepped into a spacecraft. During the Spacelab Life Sciences mission, Tamar Jernigan performs a mock medical examination on pilot Sid Guierrez, using the Crew Medical Restraint System. It was an early design to test the feasibility of performing an emergency medical examination on a patient in space.

Until the present time, anyone who wanted or needed to talk to a crew-member in space, including a doctor, had to be in Mission Control in Houston, or, during Spacelab flights, perhaps at the Payload Operations Control Center at the Marshall Space Flight Center. With telemedicine capabilities, physicians and specialists around the world will be able to contribute to medical responses in an emergency – such as contamination of the atmosphere from a leak on the station – and the best medical expertise will be available at all times.

Protecting the health of crew-members during months-long work in space also requires the development of countermeasures to the physiological effects of microgravity, so that astronauts are healthy both in space and when they return to Earth. Overall, the data from US astronauts onboard Mir confirmed the observations made during the 1970s Skylab programme. But even in this strictly 'objective' science of physical health and well-being, cultural differences have led to differing evaluations of the appropriate measures necessary for space travel.

It is well known that both bone density and muscle mass deteriorate in space. NASA reports that Phase 1 results showed that even with the very rigorous exercise programme that was carried out onboard Mir, neither integrated muscle function nor bone density were up to standard levels upon return. American investigators learned from the Russians that if specific muscle groups were targeted by doing exercise, muscle strength could be protected in those areas, but not overall.

The Russian scientific community has evaluated its data and the performance level of their cosmonauts somewhat differently. Introducing her two papers presented at the International Astronautical Federation congress in Amsterdam in October 1999, Dr Inessa Kozlovskaya stated that the efficacy of physical exercise as a countermeasure to deconditioning during long-duration spaceflights is 'one area where there are differences of evaluation between Russia and the US.' She explained that Russian countermeasures include the selection of crew-members, training, in-flight exercise and post-flight rehabilitation. She also stated that in Russia, astronaut selection continues throughout training, to ensure that each cosmonaut is fit for the mission. In the US it is very unusual for an astronaut to be removed from a flight after he is chosen for a specific mission.

After describing the rigorous regimen of exercise required of all cosmonauts, Kozlovskaya concludes that the Russian countermeasure system has proven to be 'highly successful in fighting negative effects of weightlessness on the physiological systems of the body.' She states that 'negative changes caused by weightlessness depend on the intensity and the amount of physical exercise done during the flight,'

Figure 6.18. While there are differences of opinion on the effectiveness of exercise in mitigating the effect of weightlessness during long-duration missions, it is seen as a necessary part of maintaining crew health. For the International Space Station, medical specialists are attempting to devise additional techniques for countering the loss of bone and muscle mass in space, but the time-honoured treadmill was the major device used on Mir, and will be present on the new station. The only reason that Dr Shannon Lucid was distressed to learn that she would be spending more time onboard than planned was that she would have to do more exercise on the treadmill. (Courtesy NASA and the Russian Space Agency.)

and that cosmonauts who did not combat the effects of weightlessness successfully, did not adhere to the recommended regimen.

A second paper, prepared along with colleagues of Dr Kozlovskaya, states categorically that 'sixty-five space missions to Salyut and Mir orbital stations, of 60 to 238 days duration, have proven the Russian system of countermeasures effective for preventing or significantly diminishing almost all of the negative effects of weightlessness in long-term spaceflights.' In presenting these papers, Dr Kozlovskaya acknowledged that her conclusions differ from those of her American colleagues.

There is agreement that there will have to be methods devised to ensure that onboard exercise is more palatable for the crew. When told that her stay on Mir would be extended, Dr Shannon Lucid's response was not that she wanted to go home because she missed her family, but that she dreaded more time on the treadmill! But as Dr David Wolf, himself a medical doctor, suggests, astronauts are not being sent in to space to exercise, and so additional countermeasures should be developed.

In a paper presented at the IAF congress in Amsterdam, Canadian astronaut David Williams, MD, who directs the Johnson Space Center's space and life sciences divison, and Christopher Flynn, MD, also from Houston, considered other cross-cultural questions involved in the medical treatment of crew-members on the ISS. During the assembly phase of the ISS, overall medical authority will rest with the Multilateral Medical Policy Board, which is comprised of a single representative from each international partner, and is co-chaired by NASA and the Russian Space Agency. After assembly completion, the Chair will rotate among all of the partners.

Under the Policy Board there is a Multilateral Medical Operational Panel (MMOP), in which each international partner in the ISS is contributing to the development of common approaches to space medicine. The MMOP is responsible for developing crew medical evaluation and care standards, assigning a crew surgeon for each expedition flight, and continually evaluating medical hardware and capabilities. The problems that the MMOP is tackling include differences, for example, in how readily crew-members report illness, with Americans known to be more straightforward than their Russian colleagues. There are particular pharmacological interventions that have been developed and are administered by one country, but are not accepted in another. And in some cases, the diagnosis of a syndrome does not even exist in another nation's lexicon.

On a somewhat lighter note, there may be a battle ahead over cultural differences regarding at least one particular aspect of crew health, both mental and physical, on the ISS. NASA has never allowed astronauts to consume alcohol in space. The Russians have sent small amounts of alcoholic refreshment to orbit, for the celebration of special occasions, such as the New Year. However, according Jack Stuster, the French think it absolutely uncivilised to have to dine without a glass of wine. Without wine, the French worry, they may not be able to attract new astronauts to join the programme.

At the present time there are 176 astronauts in training at NASA's Johnson Space Center in Houston. These include five from the Russian Space Agency, 13 from the

European Space Agency, six from Canada, three from the National Space Development Agency of Japan, and one from the Brazilian Space Agency. Over the course of completing the ISS it is hoped that additional nations will join in this grand project, each bringing with them the dreams and aspirations of their country, and their history, language and culture.

6.9 THE LEGACY OF MIR

During the course of Mir's operation in space, more than 100 people passed through its portals. The occasion of the final long-duration cosmonaut crew leaving Mir on 28 August 1999 prompted Mir's Mission Control director, former cosmonaut Solovyov, to reflect on its accomplishments. Commenting on the focus in the media on Mir's equipment failures during the past three years, Solovyov said: 'Why should we talk about these dramatic moments? During these 13 years, at least two-thirds of all we've done and accomplished, we never planned to do. We never thought that the American Shuttle would visit the station, for example... We never thought about this at the beginning of the 1980s when we created the station.'

Solovyov stressed that what has been accomplished on Mir is not just to the

Figure 6.19. The Russian Mir station leaves a great legacy in scientific accomplishment, and experience of living and working in space. The Shuttle–Mir programme provided a solid foundation for the international collaboration upon which the new International Space Station is founded. As Dr Andy Thomas stated, building the ISS with the Russians would have been 'inconceivable' without having first carried out the Shuttle–Mir programme.

benefit of the nation that designed, built, launched and maintained it. 'Mir station has scientific value that now belongs to all nations,' he said. 'During these 13 years, standards have been established which are accepted as the most effective and most optimal: the module-based design, step-by-step construction, the transportation system to deliver people to the station and bring them back. All this served as a foundation for the International Space Station. That's why the Mir station contributed so much to space science, for the entire world. It was very useful. We had some troubles, but we had more accomplishments.'

In recognition of the importance that Mir has played in promoting the development of aerospace technology, on 7 October 1999 the Francis-Xavier Bagnound Foundation announced that five Russian scientists who helped develop Mir will share the fourth $250,000 Francis-Xavier Bagnound Aerospace Prize. The recipients, Yuri Semenov, Anatoly Kiselev, Gai Severin, Anatoly Gregoriev and Peter Klimuk, were presented with the award at London's Aeronautical Society on 15 October 1999.

Mir was a technological achievement that functioned well beyond its rated lifetime, largely because of the competence, dedication and courage of the cosmonauts and mission managers who worked on the programme. It benefitted from a quarter of a century of Soviet and Russian experience in the design, construction and operation of space stations. Now, the International Space Station will gain the benefit of 13 years of experience with Mir.

7

Science on an international space laboratory

> One way we will ultimately overcome the economic problem associated with medical care is to obtain the knowledge needed to prevent diseases and find new means to treat patients, especially as our population ages. We cannot always predict the outcome of scientific activity, especially efforts as board and untried as space. One reason some scientists and political leaders question the efficacy of space research is that we have had limited opportunity for multiple experimentations and trial runs in space. Significant return on science research requires an ability to acquire information in both quantity and quality. Present technology on the Shuttle allows for stays in space of only about two weeks. We do not limit medical researchers to only a few hours in the laboratory and expect cures for cancer. We need much longer missions in space – in months to years – to obtain research results that may lead to the development of new knowledge and breakthroughs.
>
> Michael E. DeBakey, MD
> Testimony before Congress, 22 June 1993

Since the dawn of the Space Age, men have known that their move off this planet and into space would require infrastructure in Earth orbit to support their journeys through the Solar System. Just as the railroad stations with their small towns served as transportation hubs for the westward expansion of the United States, a station in space will be a transport link to more distant venues, a waystation for travellers on their way somewhere else, a first attempt to learn how to live in this strange new environment, and the beginning of a new form of society that will flourish and grow in the future. NASA describes the International Space Station as the first 'city in space'. From that city will go forth explorers to the Moon and to Mars.

To the early space age pioneers there was never any question that mankind would move beyond orbiting the Earth to living in space. In early 1952 Dr Wernher von Braun, who brought more than 100 of the top rocket specialists in the world from Germany to the United States after World War II, considered the development of a space station to be 'as inevitable as the rising of the Sun. Man has already poked his nose into space, and he is not likely to pull it back.' Von Braun visualised a grand station, accommodating hundreds of people. 'From this platform, a trip to the Moon itself will be just a step, as scientists reckon distance in space,' he said. Once in Earth orbit, you were more than half-way to anywhere else, in terms of effort.

Before humanity moves civilisation further away from Earth than 250 miles above its surface, a breakthrough in understanding the deleterious effects and potential advantages of the unique space environment will be required. The primary goal of the series of laboratories that the International Space Station will make available to scientific researchers around the world is to provide the tools to achieve this breakthrough.

Along the way, new products will be developed that rely on the microgravity of space. Our understanding of the fundamental principles of life and of inorganic nature will be revolutionised, as we can observe them without the masking effects of gravity, and mankind will embark on an international human experiment that functions as a melting pot for the cultures around the world.

7.1 ASSEMBLY REQUIRED

The International Space Station is the greatest technical, diplomatic and creative challenge that mankind has undertaken for the twenty-first century. Nothing comparable has been attempted, and there is nothing comparable in terms of what it will take to accomplish it. This project brings together 16 different nations, so far, in a long-term commitment, requiring significant resources from each country, each of which must stay the course while weathering political and economic storms over a period of decades. Completing the ISS will require that peoples transcend cultural, historical, and political differences, to defend the common aims of mankind in this universal project of science, technology and exploration.

The multi-year construction of the ISS will depend upon the unwavering support, politically and technically, of all of the space agencies involved. Each will have to be determined to organise political and budgetary support, and persist through technical glitches, missed deadlines, cost overruns, failures and inevitable criticism. There is no possibility that a project of this size and complexity will be completed without failures along the way.

The technical, and simply physical, challenges of the space station are truly staggering. Requiring at least 37 more Space Shuttle assembly flights, (the total number of flights from all international partners exceeding 90), and weighing more than 450,000 kilograms, the ISS will take four more years to complete. While Mir contained and enclosed a volume of about 500 cubic metres, ISS will have an internal volume more than three times that of Mir. More than 900 hours of space-walks by astronauts and cosmonauts outside the station will be needed to attach and activate the modules.

When complete, the ISS will be comprised of six scientific laboratories, in addition to the logistics, operations and storage facilities required. Two laboratories will be contributed by the US, one each from Europe and Japan, and a planned two

Figure 7.1. When the International Space Station is complete a few years from now, it will be comprised of a series of laboratories provided by numerous countries, housing astronauts from different lands, who will be laying the groundwork for even more ambitious expeditions in space.

from Russia. The US is also planning to construct a Habitation Module, with expanded crew quarters that will bring the station up to its full complement of seven, as the last major component scheduled for delivery. Over time, four huge American-built solar arrays will be attached to the station, each nearly 100 metres long, bringing the total delivered electrical power to the station to 110 kW.

The station will be serviced by a multitude of different spacecraft. The Space Shuttle will deliver components, supplies and crew-members, and will also bring back precious scientific samples and products, and hardware that requires repair or upgrades. The Russians are committed to launch Soyuz spacecraft to the ISS, to deliver crews and remain attached to the station to provide a crew-return capability. Unmanned Proton launchers will deliver the major Russian components, and unmanned Progress ships will deliver reboost fuel and dry cargo.

The Japanese plan to eventually use their H-IIA launch vehicle to boost their Hope Transfer Vehicle to the station. This unmanned supply ship will have a payload weight capability of 6–7 tons, and Japan expects to launch a demonstrator vehicle in the summer of 2002. It is planned that the Hope Transfer Vehicle would make trips to the ISS twice each year. The Europeans will use their Ariane 5 rocket for delivering their unmanned Automated Transfer Vehicle. In addition to delivering

Figure 7.2. Assembly of the ISS will require more than 900 hours of space walks to attach and activate new modules and equipment. Astronaut Jerry Ross, a member of the STS-88 mission, performed one of three EVAs which were conducted in December 1998 to secure the first two elements of the space station.

dry cargo, the ATV will carry propellant, and can reboost the station, which will tend to lose altitude due to drag in the very thin orbital atmosphere. A Crew Return Vehicle that can accommodate the full complement of seven crew-members, under design and development in the US, will later be added to the transport capabilities for the station. A grand ballet will occur in Earth orbit, as the ISS is visited and

Figure 7.3. The Russian-built Functional Cargo Block, Zarya, was the first element of the International Space Station, and was launched in November 1998. It was produced at the Khrunichev State Research and Production Centre in Moscow. Here, technicians are working on the almost completed aft portion, which includes the docking mechanism that will be used for the second Russian element.

serviced by an international array of visiting ships, requiring the greatest degree of coordination and scheduling.

The first element of the ISS was launched by Russia on 20 November 1998. The Zarya (Sunrise) control module, weighing in at more than 20 tons, provides the station's initial propulsion capability, to keep it in the proper orbit, and has solar arrays to provide an average of 3 kW of power. It is a fairly simple initial anchor for building the station, and many of its functions will be replaced by the Russian Service Module, Zvezda (Star).

Two weeks after Zarya arrived, the six-sided Unity connecting mode was brought to orbit by the Space Shuttle *Endeavour*. During a 12-day mission, astronauts made the connections, and the first two ISS modules became one spacecraft. Unity is the connecting link, with six docking ports, to which future modules of the station will be attached. The Russian Service Module, Zvezda, provides living quarters for the first crews, and the core of the early station. It is similar in design to the core module of Mir, and would have been the centerpiece of a Russian Mir 2, had that been built. It will also be able to support a limited number of scientific experiments.

The first expedition crew, scheduled for the year 2000, will be commanded by astronaut Bill Shepherd. Cosmonaut Yuri Gidzenko will be the Soyuz commander, and Sergei Krikalev the flight engineer. Devising a balance of three crew-members

Figure 7.4. The second ISS element, which was launched the month following Zarya, was the American Node 1, Unity. Members of the STS-88 crew, who delivered it to orbit, posed with the node at the Kennedy Space Center on 26 June 1997. Unity is designed with six hatches to serve as docking ports for future modules, and will function as a passageway between the living and working area of the station.

from two nations, one of whom has to be in command of the mission, was not easy to negotiate. Astronaut Shepherd will command the ISS mission, and cosmonaut Gidzenko will be in command of the travel to the station in the Russian Soyuz vehicle. Although two of the three modules will be Russian modules when the crew arrives, Zarya was paid for by the US, and with its contribution of half the entire cost and weight of the facility, the American space programme will retain the greatest degree of control of its operations.

While NASA and the Russian Space Agency have preliminary dates outlined for the remaining launches to the ISS, these will change. The down-time for Shuttle flights in the second half of 1999, while an investigation was being conducted into wiring problems in the orbiters, is one recent example of the difficulty in keeping to exact dates. On 27 October the second Russian Proton rocket in four months failed soon after take-off, these two failures having been the first, in 15 years, of the rocket itself. The Proton is the booster that will be used to deploy the Russian Service Module, Zvezda, and it is hoped that the recent failures of the Proton will not delay that launch.

Figure 7.5. Zarya and Unity mated together to form the beginnings of the International Space Station. This photograph was taken by the Space Shuttle crew on 13 December 1998, after the astronauts had completed their assembly mission and were heading back to Earth.

Figure 7.6. The Russian Service Module, Zvezda ('Star'), will be the early cornerstone of the International Space Station for human habitation, life support, guidance and navigation, and propulsion. The forward portion of the spacecraft (out of view in this photograph) has a spherical transfer compartment with three docking ports. This photograph was taken at the Khrunichev facility in May 1998.

What will not change, except under catastrophic circumstances, is the order in which the rest of the station's facilities will be launched, as these must follow a logical sequence.

After the first long-term expedition crew is on orbit, the US laboratory module, Destiny, will arrive at the station. Then the research that the scientific community has been waiting for will get underway. Soon after, the first of four giant solar arrays will arrive, as will the Canadian robotic arm, for assembly assignments. During the following three years, the other structural, operations, and research facilities will arrive.

Not only is the design of the ISS modular, so are the facilities for science. In order to provide experiment space for any nation, International Standard Payload Racks – the size of refrigerators, with experiment drawers one metre wide and two metres tall – will hold the laboratory equipment and experiments. They are modular, and are the infrastructure provided so that individual experiments can be plugged in and changed whenever necessary. They are designed for easy maintenance by the crew. The racks can be used in the US laboratory, the Japanese Experiment Module (recently renamed Kibo, (Hope)), the European Columbus Orbital Facility, and the Centrifuge Accomodation Module. In the latest design there will be a total of 33 rack spaces on the ISS.

Dedicated racks that house long-term or permanent experiments of a single discipline are called 'facilities' (such as the Gravitational Biology Facility). It is fully

Figure 7.7. When the US laboratory Destiny is mated to the International Space Station it will provide a world-class working facility for astronaut–researchers who will be living in space. This photograph, taken in December 1997, shows the laboratory frame. The exterior will be covered with a debris shield blanket made of material similar to that used in bullet-proof vests.

expected that such a rack, or series of racks, will be used for continuing experiments throughout the life of the station. Experiments that may be more temporary and only need part of one refrigerator-sized rack will be housed in EXPRESS racks (Expedite the Processing of Experiments to Space Station). Experiments in the EXPRESS racks can be easily changed.

7.2 A DOCTOR IN THE HOUSE

The facilities that the ISS will house for scientists to conduct research will be deployed piecemeal, following the procedure of assembling the station itself. The number and variety of experiments that the crew can execute at any one time will depend not only upon the facilities on the station, but also on the amount of time available for science. During the next few years, astronauts and cosmonauts will be most focused on assembling the components as they are brought to orbit, and on preparing the facilities for use. But science will be underway from the moment that the first expedition crew inhabits the station in the year 2000.

As with all previous manned space missions, the basic physiological functions of the crew will be measured in space and monitored by medical officers on the station and physicians on Earth. The Crew Health Care System will accompany the first expedition crew to the ISS to perform the basic monitoring of the health and adaptation of the crew. After the US laboratory Destiny is in orbit, more sophisticated instruments will provide continuing research data, deepening our understanding of how the human body adjusts to microgravity. For the ISS, NASA and its international partners have developed a new generation of remote-monitoring technologies to ensure that the crew on board is healthy, and to provide the best available medical care if they are not.

Telemedicine – the practice of medicine from a distance using advanced information and communications systems – has been developed on an international scale by the US and Russia, to allow clinical consultations and medical education between academic and clinical sites, linked together through an Internet-based telemedicine testbed. This highly successful 'Spacebridge to Russia' programme, developed during the joint Shuttle–Mir operations, had been preceeded by the 'Spacebridge to Armenia' programme, used to provide medical consultation services during the earthquake in Armenia in 1988.

While general paramedic-level knowledge among the crew will be normal on every mission, there will be times when a sick or injured crew-member will require more extensive medical care. In order to make a refined diagnosis, medical specialists on Earth will need information on the state of the patient, such as heart rate and blood pressure, or results from an electrocardiogram or other diagnostic instruments. The best specialised medical care will be available through telemedicine links to help the crew on board, which will still have the responsibility for treatment.

In 1997, NASA's Lewis Research Center, in Ohio, conducted an advanced telemedicine experiment with NASA's Ames Research Center, in California, and Dr James Thomas, of the Cleveland Clinic Foundation, also in Ohio. A volunteer

Figure 7.8. In a test of the remote technology of 'telemedicine', Dr James Thomas, of the Cleveland Clinic Foundation, conducts a long-distance echocardiogram examination of a patient. He is using a real-time display of the transmitted images. On the International Space Station such technology will provide astronauts with access to the world's best medical expertise. (Courtesy Tom Trower/NASA Ames Research Center.)

'patient' underwent an echocardiogram examination at Lewis, and was remotely diagnosed by Dr Thomas, who was at Ames. The doctor viewed a real-time display of images, and reported that he was 'very pleased with the diagnostic quality of the echocardiogram.' He explained that this technique will be used on the International Space Station because it less invasive, uses less power, is small and versatile, and is not magnetic or radioactive, which simplifies its use in space. The electrocardiogram uses ultrasound to produce images of the heart, providing a motion picture of cardiac function. The technical challenge for NASA was to meet the requirement of transmitting images that are high-resolution and moving in real time.

On 3 November 1999, Stephen Wyle, President of CyberMDx Inc, described for the press a new device, the Telemedicine Instrumentation Pack, that was tested on the Space Shuttle and will be deployed on the ISS, to send a patient's medical data to a doctor data via the Internet. The commercial spin-off of the NASA-developed technology is a computerised device that can be operated by a nurse or any person with training, and can perform many of the functions of a physical examination. It can monitor blood pressure and pulse, and examine the eyes and ears. Images from the examination can be sent to the doctor in real-time video or via electronic mail. The Telemedicine Instrumentation Pack is being tested by a health clinic, Mr Wyle said, and on a cruise ship.

NASA is also conducting ground research on an automated portable intensive care unit for treatment in orbit, and the technologies of cybersurgery, making use of

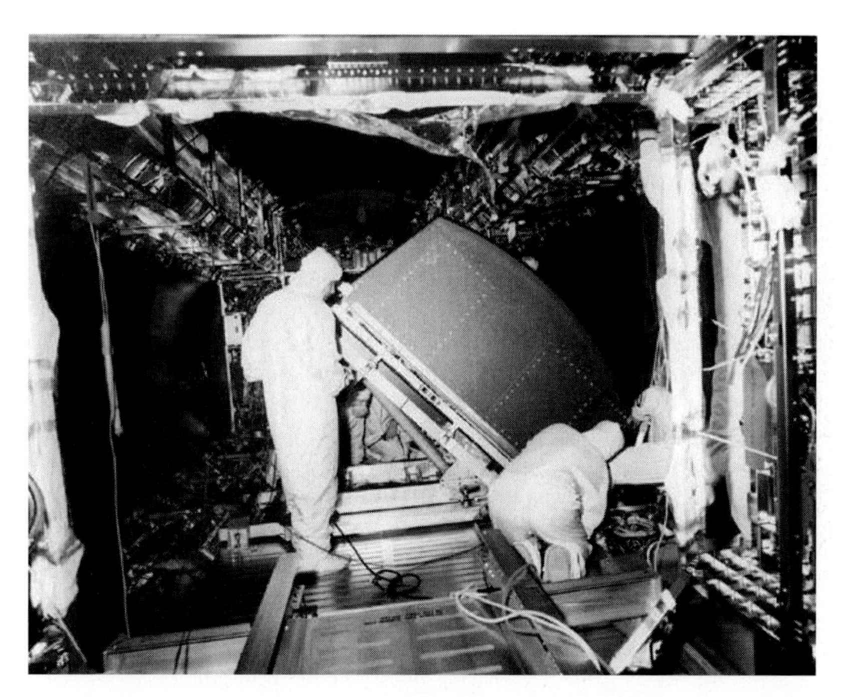

Figure 7.9. The International Space Station will house International Standard Payload Racks which will deliver the power and other resources needed for the experiments to be plugged in. On 26 March 1998, Boeing technicians installed the first rack, which will supply electrical power, inside the US laboratory Destiny. When the laboratory is complete it will consist of a total of 24 racks, 13 of which will contain experiments. The other 11 racks will provide power, temperature and humidity control, air revitalisation and other support systems.

the tools of digital models and virtual reality to assist the onboard practitioner, are under development. All of these remote technologies have applications on Earth, particularly in rural areas that lack specialised or, in some cases, any medical expertise on site, and some are already part of the rural practioner's arsenal to help maintain public health.

The Human Research Facility, under development for the ISS, will be the closest thing imaginable to a microgravity doctor's office. It will be the first facility-class payload delivered to the station. Using equipment housed in two racks, crew-members will be able to take their blood pressure automatically, monitor their heart rate, measure the gas concentration in their exhaled breath for respiration and pulmonary studies, measure the mechanical power of their arm muscles, and draw blood. They will also be able to assess the effects of microgravity on their visual system, balance, and eye, head and body coordination.

Under consideration for use in the Human Research Facility are also an activity monitor worn on the wrist to evaluate the quality of sleep, the onset of sleep, and other activities; ultrasound technology for sophisticated diagnosis, if it is needed; a hand grip dynamometer to measure hand strength; a centrifuge to separate biological samples; and instruments to collect samples of bodily fluids. In addition to providing more data on the effect of microgravity on the human body, the Human Research Facility will house the treadmill, bicycle ergometer and other exercise equipment to counter some of the delibilitating effects of space.

Not surprisingly, the Russians, who have the longest-term set of data on the physiological responses to weightlessness, are planning an extensive series of biomedical experiments in their modules of the International Space Station. In a paper presented at the 50th Congress of the International Astronautical Federation in Amsterdam in October 1999, Dr R.M. Baevsky and colleagues from the Institute for Biomedical Problems in Moscow outlined their plans for a comprehensive study of the problems of cardiovascular regulation in space. They have focused on this area because 'blood circulation is indicative of the adaptability of the whole body'. They are most concerned about the initial phase of ISS assembly, because of the 'expected heavy physical and psycho-emotional loads upon the crew-members due to [the] large amount of assembly operations to be done in orbit.'

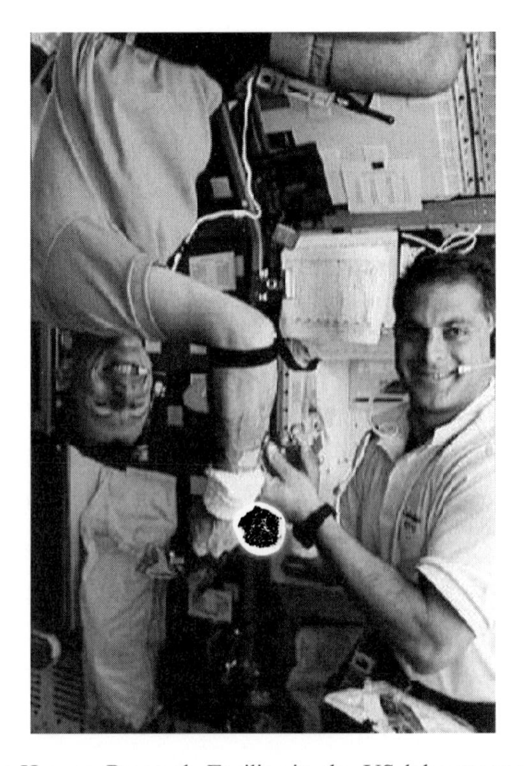

Figure 7.10. The Human Research Facility in the US laboratory will be the closest equivalent to a doctor's office on Earth. Until now, much of the medical research in space has been conducted in less specialised venues, and sometimes 'on the fly', such as this blood draw by medical doctor and astronaut David Wolf from payload specialist Martin Fettman during the October 1993 Spacelab Life Sciences mission.

The first experiment planned by Russia for implementation on the ISS is the Pulse experiment, primarily intended for the verification of data obtained during the Mir missions, which had indicated variations in cardiac rhythm. The Pneumocard cardiorespiratory experiment will study the functioning of the heart, and the oxygen concentration in the blood. Cardiopres will provide a continuous supply of data to scientists on the crew-members' blood pressure. And Sonocard will study the recovery of 'functional reserves' during the night.

The scientists report that these techniques are being tested during the eight-month isolation and confinement experiment (described in Chapter 6), which 'mimics a multinational ISS mission'. One purpose of this set of experiments is to monitor the crew for the first signs of strain during their heavy-duty assembly missions on the ISS.

The European scientific community is also contributing to crew health care and the study of the effects of microgravity on the human physiology. In 1997 the French and German space agencies agreed to jointly develop Cardiolab, to be housed in the European Columbus laboratory on the ISS. Both agencies drew on their years of experience flying biomedical payloads on Mir. While there will be some overlap with other laboratories, the goal of Cardiolab is to complement other resources. The emphasis in Cardiolab is to make use of off-the-shelf medical technology that is already on the commercial market, with modifications and upgrades.

A team of French and German researchers described some of the interesting and innovative devices planned for Cardiolab, at the International Astronautical Federation Congress in Amsterdam in 1999. One of them is a Doppler instrument to measure blood flow velocity using ultrasound pulses in up to three arteries at a time. The Applied Potential Tomography device is based on the method of imaging biological structures by measuring the electrical properties of the tissues in cross-sectional planes of the body. This is accomplished through an injection of alternating electrical current and by measuring the resulting electrical potential differences by a circular array of electrodes attached to the surface of the area.

Cardiovascular measurements on the crew-members will be taken while they are at rest and during their normal activities. The facility also contains a series of stressors, such as a warm/cold glove, a static calf ergometer to impose contractions on the calves, and an arm–leg cuff to apply stress to the limbs. The facility will be available for international researchers, and, as in the other laboratories, its modular design will allow for a flexible arrangement and variety of experiments.

7.3 CRYSTALS, CANDLES AND CREATURES

After the US Laboratory Module Destiny has been delivered to the ISS, it will be outfitted with the first of three Materials Science Research Racks. This facility will allow scientists to carry out experiments to understand the fundamental properties that control how materials form in space, to enable the design of new materials which can be directed, rather than by trial-and-error. Researchers also plan to produce small, improved high-value products, such as materials for advanced electronics for use on Earth.

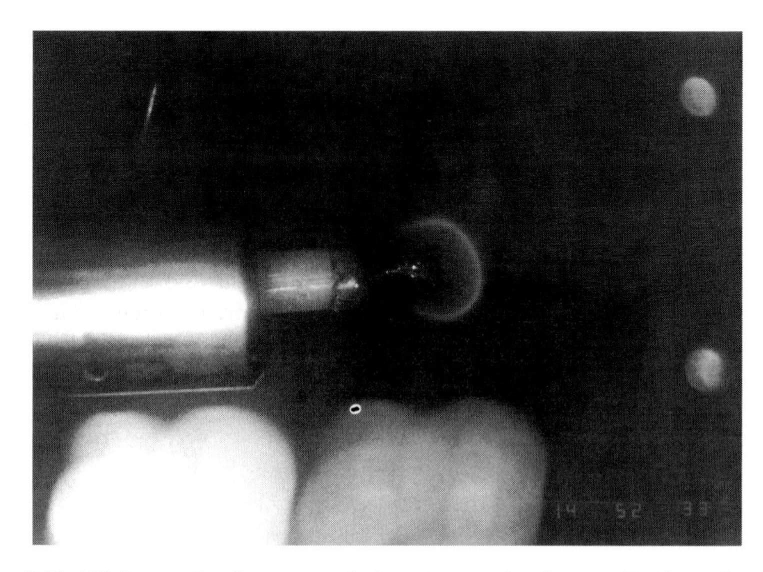

Figure 7.11. While combustion research in microgravity has applications for fuel-burning and other activities on Earth, it is of more than academic interest to space travellers, who may one day have to fight a fire in space. Experiments have demonstrated that in space a candle burns with a round flame, due to the absence of convection currents, as can be clearly seen in this image from a Space Shuttle experiment conducted in 1992.

At the Third Biennial Microgravity Materials Science Conference, held in Huntsville, Alabama, in July 1998, Dr Frank Szofran, of the Marshall Space Flight Center, reviewed the status of materials research planning for the ISS. He noted that the first Materials Science Research Rack will include an Experiment Module that will occupy half of the rack. The European Space Agency will provide the core structure of the Experiment Module, which will be designed to house five modular devices that can be inserted for individual experiments. NASA and ESA will each provide two insert modules, and the German space programme will provide the fifth, which will use a special furnace with a rotating magnetic field to control flows within molten metal samples. Each of the first five experiments will be a furnace to process materials, each using a different method, to discover the process at the moment a molten metal turns to a solid.

In a briefing for recently selected astronaut candidates, Dr Sharon Cobb, the project scientist for the Materials Science Research Facility on the ISS, reported that she is often asked why there are so many furnaces for materials processing planned for the station. 'Well, most of us have several ovens at home – a toaster oven, a microwave oven, and regular conventional oven, and then four hot eyes on the top of the oven', each for different purposes. Similarly, it is almost impossible to design one furnace that could satisfy the requirements of every experiment. Some samples need to be heated and cooled evenly, others require gradient heating where a 'hot zone' travels down the length of the sample material, and still others require a rapid quench where the material is quickly cooled to retain the history of the physical process under examination for analysis on the ground.

The other half of the first Materials Science Research Rack will be occupied for nine months by equipment from NASA's Space Products Development Division, to be made available to commercial microgravity researchers.

Experiments that need a 'personal touch' and hands-on attention will be housed in the Microgravity Science Glovebox unit, which will occupy a double-floor-to-ceiling International Standard Payload Rack and will have openings 40 cm wide to accommodate fairly large experiments. The Microgravity Science Glovebox unit allows crew-members to insert their hands into special gloves and then use them to manipulate samples that are contained in the sealed chamber – similar to the handling of radioactive materials or infectious pathogens in laboratories on Earth. It will be a very versatile facility, servicing experiments in biotechnology, combustion science, fluid physics, fundamental physics and materials science.

Like a workbench in a typical laboratory, the glovebox supports experiments with electrical power, air conditioning, a connection to the station's communication network, and color video capability. The sample can be enclosed in pressurised nitrogen, and the air flow will keep the samples clean as well as control the temperature. The Microgravity Science Glovebox, which will be placed in the US Destiny laboratory, is a contribution by the European Space Agency, which delivered a ground-test model to NASA's Marshall Space Flight Center in

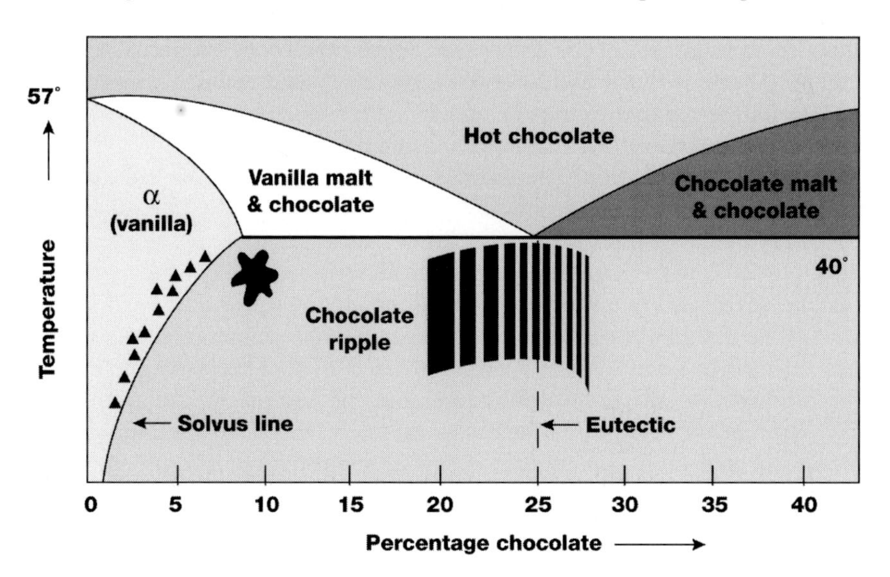

Figure 7.12. A broad range of materials science experiments will be conducted on the International Space Station, including the study of the different phases that various material pass through at different temperatures. Dr Sharon Cobb, of NASA's Marshall Space Flight Center, has developed this fanciful materials phase diagram as an heuristic device, showing how changes in temperature and composition effect change in the mixing of chocolate and vanilla.

September 1999. Using the ground unit, scientists will be able to bring in their experiment hardware and install it in the glovebox, to ensure that it is fit and ready for the real station facility.

Combustion is one of the processes that is greatly influenced by the force of gravity. On the ground, the warmer air currents near a candle flame rise, removing the heat, and allowing the flow of oxygen to feed the flame as cooler, dense air rushes in to replace the warm air. But in space, rather than forming the familiar conical shape that we see on Earth, the flame of a candle is spherical. Without being distracted by the effects of the air flow around a candle, in space, scientists can study the process of combustion itself. The US Fluids and Combustion Facility includes three racks, two of which will support a range of experiments in combustion, using gaseous, liquid and solid fuels.

The understanding of combustion in space has one rather obvious application: the control of fire onboard a spacecraft. But in addition to applying that knowledge to fighting fires on Earth, combustion of fossil fuels accounts for 85% of the world's energy production. Investigations will include hardware to test nozzles for burning gases, and other techniques to improve burning efficiency. It is hoped that improvements in the combustion process can be found, which would, even if minor, have a dramatic effect in increasing the efficiency of energy production on Earth.

The third rack will contain a fluid module with several experiment test chambers. Fluid physicists are interested in studying the surface tension effects, boiling mechanisms, directional solidification of liquids and other aspects of the behaviour of fluids in microgravity. The European Space Agency is planning to outfit its Columbus module with a Fluid Sciences Laboratory with similar research goals.

An aspect of space that cannot be matched anywhere else is the view. Most crew-members have looked through the windows of spacecraft not only for the study of the Earth but for relaxation. On the ISS, crew-members will be able to view the Earth from 51°.6 above to 51°.6 below the equator, covering 75% of the Earth's surface, where 95% of the Earth's population lives.

The Optical Window Rack Facility on the ISS will be made of the highest optical quality glass ever used in a spacecraft. It will be mounted within the US laboratory. A rack will be placed above the window to accomodate cameras, sensors and other devices. Instruments sensitive to a different parts of the electromagnetic spectrum will be calibrated at the Optical Window, and the astronauts can reconfigure the available equipment to capture unexpected events on Earth. It will undoubtedly be a most popular spot on the space station, for observation and enjoyment.

Gravitational biology is the study of gravity's influence on the development, growth and internal processes of plants and animals. On the space station, the long duration will present the opportunity to carry out multigenerational studies of living organisms. Because of the modular nature of the station's science facilities, scientists will also be able to pursue the study of gravitational effects on a wide range of samples, including cells, tissue cultures, full-size plants, insects, rodents, aquatic specimen and avian and reptilian eggs. The goal is to study the role of gravity from gestation through ageing. The American Gravitational Biology Facility is a two-rack facility with interchangeable habitats, depending upon the subjects for study. The six

Figure 7.13. Gravitational biology will take a great step forward when different biological specimens can be subjected to partial gravity forces through the use of a large centrifuge on the station. Smaller centrifuges have flown in space, and in this 1995 image astronauts Leroy Chiao (top) and Donald Thomas are at work in the Space Shuttle. Chiao is placing samples in a centrifuge in a rack, and Thomas has his hands in the 'glovebox' where samples for the slow rotating centrifuge are prepared.

specialised habitats which are planned for the facility each provide food, water, light, air, humidity control, temperature control and waste management for the creatures living within it. The habitats will also collect science and engineering data and transmit it to the investigators.

The Advanced Animal Habitat is a residence that can accommodate a dozen mice or up to six rats. A Mouse Development Insert can be used which will accomodate pregnant mice and can house offspring from birth through weaning. The Cell Culture Unit will be used for cell and tissue biology research. The Aquatic Habitat, being developed by the Japanese space agency, can accommodate small freshwater fish for up to 90 days, and egg-to-egg generational studies to examine life at all stages of development. The Plant Research Unit will support plants up to 38 cm in height, and the Insect Habitat, being developed by the Canadian Space Agency, will support both multigenerational and radiation studies on species such as the universally studied *drosophila melanagoster*, the fruit fly. An egg incubator will care for small reptilian and avian eggs through hatching. It is expected that the first element of the laboratory will be launched on the third utilisation flight, with the final element currently scheduled for launch near the end of the assembly sequence in 2004.

The European Space Agency is developing the Modular Cultivation System, which will be housed in Destiny, the US laboratory. It will provide a long-term growth facility for plants, to study the multigenerational effect of gravity. Experiments with insects and amphibia will be conducted, as well as cell and tissue culture studies. The Modular Cultivation System will provide a thermal control system incubator, an atmospheric control system and two rotary centrifuges. For similar studies, Japan is developing the Cell Biology Experiment Facility, and the US is developing a Plant Research Unit and Cell Culture Unit.

For its own Columbus laboratory, ESA is building Biolab, which will also include centrifuges to study the variable gravity (0–1 g) effects on cell cultures, small plants, microorganisms and small invertebrates. Biolab will be launched as part of the initial payload inside Columbus, scheduled for 2004. There are two sections in Biolab, for automated and manual control. Without crew interaction, experiment samples can be incubated, stored at cool or cold temperatures, prepared for study, and microscopically or photometrically analysed.

The aim of gravitational biology research is to not only study the effect of microgravity on living creatures, but to study *variable* gravity effects. If man is to live on the Moon or Mars, he will be subject to some gravitational force, although less than that of the Earth. Will 'some' gravity be as good as Earth's gravity, in preserving the vital functioning of man? We will not know until we can measure the effects of partial gravity on living things.

A large centrifuge for the International Space Station is being developed by the National Space Development Agency of Japan, as part of Japan's payment in kind to NASA for launching its Kibo Experiment Module on the Space Shuttle. NASDA has contracted the Toshiba Corporation to build the 2.5-metre centrifuge, which will be housed in its own Centrifuge Accommodation Module. The centrifuge will spin at adjustable speeds, providing artificial gravity (centrifugal force) in a range between 0.01 and 2.0 (double the force of gravity on Earth). By keeping all variables except the force of gravity constant during an experiment, scientists will be able to more precisely isolate the effects of gravity on organisms from other factors.

The centrifuge is large enough to accomodate eight habitats from the Gravitational Biology Facility, and two different gravity levels can be used simultaneously by placing experiments at different distances from the centre. The sophisticated design will even allow for the refilling of water containers without stopping the centrifuge.

It has been proposed that using intermittent exposure to artificial gravity may be a useful countermeasure to the debilitating effects of the space environment, especially on long space-journeys. But scientists do not even know today whether or not there is a threshold gravitational force that is necessary to retain an Earth-compatible physical condition. We will not suggest that space explorers go to other parts of the Solar System, only to find they cannot return to the rigours of 1 g on Earth. These kinds of question will be on the agenda for research with the centrifuge.

Once various subjects have been exposed to variable gravity in the centrifuge, or other conditions of experiments, some will be sacrificed on orbit so that scientists can study the gravity effects on cells and tissue on a variety of living beings. A key workbench facility for the gravitational biology and overall life sciences research on the station will be a Life Sciences Glovebox, for operations as delicate as surgery on specimens. It will be able to accommodate operations using materials that can be hazardous to the crew, such as chemicals and radioactive materials, and will allow simultaneous use by two crew-members. This facility is also being developed by the Japanese space agency, and it is scheduled for launch on the third utilisation flight to the ISS.

There will be experiments on the outside of the International Space Station, as well as in the laboratories. The Japanese Experiment Module Exposed Facility is an

Figure 7.14. Scientific experiments will not only be mounted inside the station's laboratory modules, but will also be attached to the outside to be exposed directly to the space environment. Nobel Laureate Sam Ting has designed the Alpha Magnetic Spectrometer to search the heavens for antimatter, by sifting through the cosmic rays that will bombard the instrument as it sits outside the station. The prototype of the instrument, which flew on the Space Shuttle in 1998, is pictured here at the Kennedy Space Center before being placed inside the Shuttle's payload bay.

exposed platform, or 'back porch' attached to the laboratory, which has ten mounting spaces for experiments. It will include a Low Temperature Microgravity Physics Facility, which will be completely automated, and attached to the station externally. It is a liquid helium dewar which will study the effects of the extreme cold of only about two degrees above absolute zero.

The most ambitious external payload that will adorn the ISS is the Alpha Magnetic Spectrometer, proposed by Nobel Laureate Sam Ting, of the Massachusetts Institute of Technology. Dr Ting intends to search for evidence of antimatter by sifting through the cosmic rays that will bombard the outside of the spacecraft. A prototype of the AMS flew on the Space Shuttle mission STS-91 in June 1998, to test the basic design. This preliminary flight was necessary because the ISS version, which will spend four years outside the ISS, is a self-contained experiment that cannot be repaired by the station crew.

7.4 A REAL JUGGLING ACT

In a presentation to the International Astronautical Federation Congress in Amsterdam in October 1999, Dr Laurence Young, Director of NASA's National Space Biomedical Research Institute, and a researcher at the Massachusetts Institute of Technology, painted a graphic picture of the conflicting interests that will have to be reconciled to make scientific work on the International Space Station productive and successful.

In carrying out the scientific research, Dr Young stated, it is not 'national differences' that will be a problem. Most researchers already operate in an international environment, and most concepts in science are universally understood. But, he cautioned, there will be professional and cultural differences that 'may be critical'. For example, there is the astronaut culture. From the standpoint of the crew, he explained, it is most important to be involved in the experiments, as we have heard from Dr Lucid and the other Mir astronauts. The astronaut, Dr Young said, wants to please the Principal Investigator, and does not want to make any mistakes. If there is a problem, crew-members want to fix it themselves, rather than tell Mission Control. Ultimately, he said, the astronaut wants to satisfy the 'real boss', the head of the Astronaut Office at the Johnson Space Center, who makes the decisions about when and if he will fly again.

The crew-member, who will also be a test subject in the human life sciences experiments, Dr Laurence continued, wants to limit the amount of time he has to spend in training for a mission. He is concerned about risk from the experiments, and his privacy when it comes to the data produced. Naturally, he is also interested in minimising discomfort during the human life sciences experiments.

NASA management personnel, responsible for the entire operation of the man-in-space programme, see it in their interest to reduce the time that astronauts must allocate for experiments, reduce the training time for the experiments, simplify the procedures for the astronaut, and avoid making any changes in procedures once in space.

Then comes the Principal Investigator – the scientist who has spent years of his professional career preparing the experiment that the astronaut is entrusted to carry out in space. He wants maximal crew training for his experiment, and would actually prefer to run the experiment himself from his laboratory, giving the astronauts virtually no control over it. When it comes to the experiments where the astronaut is a subject, the PI would like to involve the maximum number of subjects, and assure that there is full disclosure of all the data, counter to the wishes of the astronauts.

Although Dr Young did not mention it, there is also the tension that arises *between* Principal Investigators, who must 'compete' with one another for space and crew time on the station. In many cases a scientist considers the research he is carrying out more important than that of colleagues, especially those in a different scientific discipline.

The flight surgeon, Dr Young said, who is part of the crew-member's ground support team, generally does (and should) take the part of the astronaut in any discussion. He needs to assure the health and safety of the crew, which includes maintaining their privacy. As a result, Dr Young reported, the flight surgeon generally 'keeps at arm's length from the scientist.'

Those at the Johnson Space Center who train the astronauts have yet a different set of criteria, Dr Young stated. They want to teach the crew 'what to do, not why'. They feel it is their job to 'interpret [for the crew] what the Principal Investigator *really* meant!'

It will take patience, open-mindedness and flexibility to have a space station running smoothly. NASA is well aware of the sensitivity that will be necessary for these various concerns.

7.5 BALANCING ASTRONAUTS AND BUTTONS

The astronauts – especially most of those who spent time on Mir and have a scientific background – emphasised that they would want to be intellectually engaged with the experiments, and the investigators during long-term expeditions on the ISS. I asked Dr John Uri, currently ISS Lead Increment Scientist, and Mission Scientist during the Phase I Shuttle–Mir programme, how a planner such as himself balances this concern with the desire on the part of some of the scientists, and NASA management, to have the experiments be as autonomous as possible, or even controlled from the ground, to help manage the crew's time. 'I think you do a little bit of both,' was his reasonable reply, 'given the limitations of how much a Shuttle can carry, and that we are only getting a piece of that [cargo space for scientific equipment] because we're launching hardware to assemble the station. There ought to be a mix. There ought to be some [experiments] that are operated from the ground, because the crew time is limited. Early on, we were told, crew time will be very limited. That doesn't mean that you can't fly those things that are challenging to the crew. We will be doing some of that in the cell science area from Neal Pellis' group at the Johnson Space Center. We'll eventually be doing some plant growth experiments – the kind of things that take a lot of care and feeding, not just pushing buttons, but where a crew-member has to look at a situation, and analyse it.'

From the very beginning of the expedition increments, Dr Uri said, there will be a mix of experiments. Some will be self-contained and more or less automated, and 'that require the crew to check on them every day to make sure the hardware is working fine, that the filters are cleaned so they're not getting clogged up, and so on. And other ones will be more hands-on kind of experiments that the crew can get involved in.' He also pointed out another constraint on crew time, and that is not in space, but in training. It is important to maximize the use of the limited resources of crew time, he said, 'and not so much crew time on orbit, but actually getting crew-member training time on the ground because they are so busy with training here and in Russia – training on Russian systems like the Soyuz and Service Module, and training on Shuttle systems, training on the US laboratory, and the [Unity] node. Their training time is very limited, also with going back and forth between here and Russia. Each time you do that, you lose a day or two, just in travel time. It is very challenging, just finding enough time for these guys to get trained to the proper level, both on systems and on the payloads.'

The astronauts have made recommendations from their Mir and Shuttle experiences on the importance of communications with scientists on the ground, in order to make the most efficient use of their time on obit. Dr Uri agreed that such availability of the Principal Investigators was critical, and that one thing NASA observed from the Shuttle–Mir missions was that 'the Russian system was actually more flexible than our previous experiences on Spacelab missions on the Shuttle.' The Russians do not restrict the number of people that the cosmonauts can talk to, and allow their investigators to talk to the crew as needed, 'given the limited amount of communications available on Mir'. The wider of range communications was also enjoyed by the crew, just for the sake of variety.

'I know that during our missions on Mir,' Dr Uri reported, 'our crew-members talked to not just our operations lead, who served as our capcomm, but they talked to a variety of our support folks, in the course of a four or five-month mission, and they appreciated it. They didn't want to hear the same voice, day in and day out.' This is certainly important in terms of working with the experiments on board. 'You want to get the right person with the right answers. You don't want to go through playing [the game of] 'telephone', passing it back and forth, where the crew-members talk to the operations leader, who passes it to the science support leader, who then passes it to somebody else, before it gets to the PI. The [crew-member's] question may get garbled. When the answer comes back, it may be too late, or they may be answering the wrong question.' Dr Uri assures that 'we're definitely anticipating that. In some of the early simulations we've had already, we've talked about that issue.'

Can there be too much of a good thing? I asked, recalling the observations of Dr Lucid when she was a capcomm. 'You don't want to do it every day,' Dr Uri responded. 'You don't want it to be a free-for-all, but when there's a requirement and the crew says, "I think the person I really want to talk to is the PI..." – or it may be the person who knows the hardware the best, who is not always the PI – we ought to make sure that the communications loop is available. Technologically, it's not a hard thing to do. It's almost more of a philosophical shift away from the way we've been doing business before.'

The same issue arises in terms of communication with operational support personnel on the ground. 'We're trying to [devise a plan to] keep from interrupting the crew with a jabber of questions, or status reports, and 'gee, how are you doing with this?' and, 'what's happening with that?' That does get annoying after a while. Our general philosophy is, don't call the crew unless they've asked a question and you're coming back with an answer. Let them ask the questions, or give us the status report, and then at the end of the day we might ask a few quick questions, such as 'what did you guys get done today? any problems? any issues? Do we need to replan tomorrow because of something that happened today?' '

Another issue that the astronauts have raised from their experience on Mir is the question of balancing their desire for autonomy and flexibility in the schedule with the need for guidance from the ground. Dr Uri commented that there has to be a way to balance both requirements. 'It's not as loose as 'here's your three-month flight and this is what we want you to get done before you come home.' Because that's not going to work. The crew-members can't, by themselves, manage all of the resources that are required. They can probably manage their time fairly well, but they don't know the parameters, such as the power situation, as well as people on the ground. It also depends upon the experiments. Some are more flexible in scheduling than others.'

Dr Uri is optimistic that the ISS will operate more like a laboratory than like a spacecraft, at some point. But during the assembly phase, 'we won't have the luxury of having crew-members that are dedicated to doing experiments. They'll be doing a lot of the assembly tasks, EVAs, and the experiments. We're not going to get there for a while, until we have a larger crew, more modules, and facilities on board. We're probably three, four, five years away from that. We've heard from Shannon Lucid and Dave Wolf that that's really what we ought to be doing, as much as possible.'

Concerning the strictness of the schedule, Dr Uri explained that 'different experiments have different requirements for how they can be timelined. For some it can be 'between now and the end of the mission', and it really doesn't matter when you do it. You just want to do it when you're not interfering with somebody else. We do have a mission timeline, like we did for Spacelab and we did for Mir, but I always regard that more as a guideline, because if you build a timeline today, and the last day of the mission is four or five months from now, I guarantee you that's not what you're going to be doing on that day. Things are going to change in the meantime.'

He provided examples of when there is virtually no flexibility on an experiment that is scheduled to be carried out. 'There is one experiment that looks at crew-members before and after an EVA. It looks at lung capacity, so obviously you have to schedule those around an EVA. Some of the experiments require exercise, so you have to schedule them during an exercise session.'

NASA has devised a method of producing a three-week planning cycle, and 'each week, you talk to the crew' and outline the activities to be accomplished. Some of those experiments are flexible, because they will depend upon circumstances beyond the crew's (or the ground's) control. 'If you're going to take some pictures of a site on the ground, this has to be done when you're going over it. But if there is cloud cover, then there's another chance the next day, and you try to schedule around that.

So when they get their daily plans, it will have some things that have to be done at a certain time, for example, because you have to do it right before you eat, or you have to wait at least an hour after eating, and those activities are put in the timeline with the time you need to do them.'

The crew will be able to carry out other tasks whenever they have the time. 'It's a combination of relatively strict timelining, as we had on Spacelab, versus the looser timelining we used on Mir, and the Russians typically used on Mir. I think that's a good balance. It gives the crew some autonomy, and gives the ground some level of control over what's going on.' And at the end of the day, the crew will report what was accomplished. 'The daily plan', Dr Uri explained, 'is essentially derived from the top-level mission timeline, and then we have the weekly or three-week plan, and get more and more focused down to the day, giving the crew as much flexibility as we can.'

The watchword over a multi-month space mission clearly should be 'expect the unexpected'. For instance, if a volcano were to erupt, NASA would hope that the crew would be able to obtain a photograph of it. You'd have to 'drop what you're doing' when the station is over the volcano, and 'then you've got to rearrange your day.'

Dr Uri is impressed with the way in which the Russians have tried to 'minimise the separation' between the way the crew carries out its work when they are on Earth, and when they are in space. 'I think the more home-like you can make the environment, and the more home-like you can make the schedule, the better off everybody will be. In most labs I've been in, you typically get in in the morning, and do a quick status check on how everybody's doing, and what you're going to do that day, or that week. You have a staff meeting. That's how I envision the work on the ISS.'

A critical activity of people who work in a laboratory on Earth is to return home at night. Cognisant of the psychological stresses that erratic crew schedules have produced in past missions, leading to reduced effectiveness in the work, the international medical oversight organisations for the ISS have set out how a 'nominal' day on orbit should be organised. Drs Williams, Nicogossian, and Shepanek from NASA headquarters presented the results of their deliberations at the International Astronautical Federation Congress in Amsterdam in October 1999.

The crew work-day will consist of 8.5 hours of sleep, 1.5 hours post-sleep (personal hygiene, breakfast, and so on), 0.5 hour planning and coordination, 8 hours of system and payload operations, 1.5 hours for a midday meal, 2 hours of exercise and 2 hours of pre-sleep free time.

The authors stress that 'space is, unquestionably, the most extreme, demanding, and unforgiving environment into which humans have ventured. The translation of medical care to space is one of the greatest challenges physicians have ever faced.'

7.6 WHAT YOU DO WHEN YOU GET THERE

Dr Uri, ISS Lead Increment Scientist, is working – within the framework of the international working groups that have been established to coordinate the scientific research on the ISS – to deal with the concerns of the different 'cultures' that will be operative during its lifetime.

On the question of finding a balance between the various scientific disciplines that are represented, Dr Uri explained that 'there are a couple of different levels where that comes in. One, there's a multilateral agreement among all the international partners about how much resource each partner gets. That's a time-averaged level of resources, not necessarily for any given increment, but over a year or two years.' Then, each partner will decide how it is going to use those resources.

The Research Planning Working Group has been established by the international partners in order to perform the 'delicate balancing act of integrating research plans for each discipline on the ISS', NASA managers from the Johnson Space Center reported in Amsterdam. The Working Group will coordinate the recommended science experiments within the physical constraints of both the ISS and the transport vehicles that carry payload to and from the station. Each partner will receive an allocation of resources on ISS thirty months before the increment is to begin. An Increment Science Team for each expedition mission will manage the balance between scientific disciplines, within the allocation of national resources for one particular mission.

There are also international planning groups within different science disciplines, to coordinate their research. For example, in 1995 the International Microgravity Strategic Planning Group was formed, with participation from the US, the European Space Agency, Japan, Canada, France and Italy. They are coordinating an

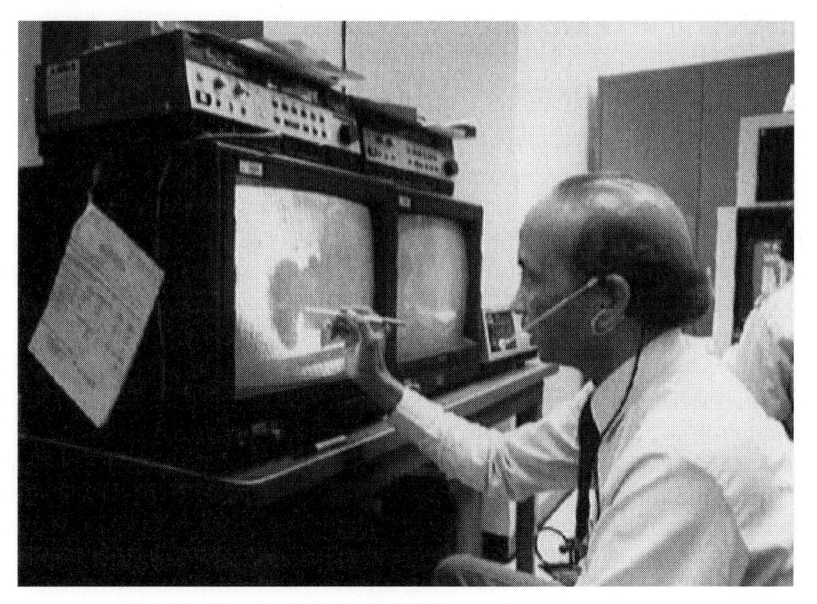

Figure 7.15. Research on the International Space Station will afford scientists on the ground – the principal investigators for onboard experiments – the opportunity to be more involved with the astronauts, who are their eyes, ears and assistants. Here, in 1985, Dr Ravinda Lal, the Principal Investigator for a crystal growth experiment on the Space Shuttle, is in the Payload Operations Control Center at the Johnson Space Center, following live downlinking of data from his experiment.

international solicitation of proposals to obtain the highest quality research for the station. In a presentation before the International Astronautical Federation Congress in 1999, Judith Robey, of the Microgravity Research Division at NASA in Washington, explained that this coordination is now more a matter of necessity than preference, because 'funding resources are an additional limitation, and are becoming the most important factor in determining the design of the [research] facility.'

Within the NASA allocation of resources, Dr Uri said, the general outline was laid out at the start of the programme. 30% of the resources will be dedicated to life sciences, a comparable 30% for microgravity sciencies, 30% for space products and development in the commercial sector, and the rest for the remainder, which includes Earth sciences, space science, and anything else that falls outside the primary categories. Dr Uri admits that 'when you get to the power, the crew time, or other constraints, it's going to be hard to equalise among all the research disciplines. How are you going to do that.'

Part of the answer is the fact that there are inherent differences among the disciplines. 'Some experiments are very power hungry, such as materials processing furnaces, which take a lot of electric power. [However,] they can be automated and run from the ground, without a lot of crew time. The biomedical studies, by nature, require crew time, because you're using the astronauts as subjects. so they are crew-intensive, but maybe not as power intensive. As a result, you cannot just insist that the 30–30–30–10 allocation will work for all resources', he explained, 'because that just doesn't make sense.'

'I think ultimately,' Dr Uri said, that based on his previous experience, 'as we go through the programme, some of that will just balance itself out, in terms of how you divide up your resources based on the specific needs of the research programmes. There's going to be a give and take. On a particular increment, you put a set of experiments together. You may have five biomedical, eight for microgravity, and so one, and when you rack it all up, it turns out that we can't do all of that because we've exceeded our power limits on that flight.'

At that point, 'you would have to go back and say, "What do we really want to do?" You may have to move this experiment to the next increment when we will have more power because new solar arrays are coming up. [Planning the science for the missions is] going to be an iterative process,' he concludes. 'That's kind of how it worked on Mir. We didn't have allocations on Mir as to who got what. It just kind of worked itself out, surprisingly, without too much gnashing of teeth.'

One reason is that 'there are always more that people want to do than there are available resources. That's going to be a fact of life, no matter where you are, including space. Then you have to be based on strategic priorities.' At the time of our interview, in August 1999, Dr Uri reported that 'we're on the middle of it right now. The first three or four increments, we've pretty much figured out what it is we want to do. We're still learning a lot about the ISS and what the real power constraints are, what it's going to take to assemble it, and how much crew time is going to be eaten up with EVAs.' The EVAs themselves are very fluid, he stated. They 'move around', because we may find that 'we thought we needed this assembly task there, but we can defer it to later, so it frees up crew time in one place, but takes it away in another.'

Even if you think you have everything well planned, he warned, 'as you know, with spaceflight it can be unpredictable.' You may have selected your experiments and investigators 'and they're all ready to fly on this particular mission, to get to orbit, and boom! Something happens, like we faced on Mir. Hopefully, we will not have anything as dramatic as the events we had on Mir, but there are always little things that go wrong, and then the crew has to pay attention to the systems. The experiments, during the assembly phase at least, will take a back seat to make sure that the station is built and working properly.'

In order to ensure that the scientific community has an active voice speaking on its behalf in the planning of the International Space Station, NASA created the position of Chief Scientist for the ISS, which is a two-year position, rotated among specialists in different fields of science. In 1998, Dr Kathy Clark, a researcher in life sciences, began her two-year term in this position, at NASA headquarters in Washington.

Dr Clark reported in an interview that 'there is no job description for the chief scientist position', so she feels she has been given 'free reign' to determine how to best serve the scientists participating in the programme. Dr Clark has taken on the responsibility of reaching out to broad layers of the scientific community to involve them in space research. She has also spent time with science writers and even Hollywood film producers, to provide advice and to develop broader support for the International Space Station.

She reports that often people not familiar with the opportunities for science in space are quite surprised at her presentations. 'Really, there aren't very many disciplines that couldn't take advantage of a little creativity, and that comes as a surprise to a lot of scientists, in general.' One focus has been to establish relationships with other scientific institutions. This, she says, has been especially important for NASA's Office of Life and Microgravity Science and Applications. Before experiments are sent to the station, Dr Clark said, 'you want to get a nice ground-based research programme going on, to get all the control [experiments] in line before you go into space, which is very expensive. There isn't any reason in the world why the National Institutes of Health couldn't fund some of that research. Then, you bring in more scientists who are interested in understanding some aspect of spaceflight.'

Dr Clark spares no frankness in describing some of the changes scientists will have to make to be able to participate in the station. There will have to be more rigorous peer review, she suggests, to maximise the good science that can be learned within the limits of the resources of the space laboratory. Sometimes, she explained, even though space research is still at the observational stage, researchers want to study the details of a particular event, even if we are not sure whether the event even exists. As an example, she pointed to an experiment that is slated to go up on the station, to investigate countermeasures to kidney stones, even though 'we've never had a reported incidence of kidney stones. I'm not sure we need to do the countermeasure yet. We need more data... The reason why [this experiment] came up was very valid. If you look at what happens to astronauts in space, and compare it with the data we have on people who get kidney stones, it would look like they could be susceptible. We really don't want that to happen. But, on the other hand, I think we need more observational data to find out if it does happen, and is really a problem. Otherwise, why deal with that?'

She cautions that it is also very important to think through the consequences of any individual experiment. 'We've talked about putting a centrifuge up there, so you could spin people once in a while [to counter the effects of microgravity, such as muscle deconditioning and bone density loss]. Unfortunately, that means that they have to readapt to zero gravity every day, and that is going to make them sick' – a reference to the space motion sickness suffered by the majority of space explorers.

7.7 SCIENCE VERSUS COMMERCE?

Another balancing act which is part of the planning and operation of the International Space Station is that between the scientists and the private corporations or other interests that the space agencies are trying to coax into using the new space facility. Such organisations will be spending their own money in preparing an experiment for flight, and intend to keep the results proprietary, as compared with the research scientist, who is anxious to publish and have access to similar research in his field. Some hostility has been voiced by scientists to the fact that 30% of the ISS's resources will be allocated to commercial concerns, who do not have to share their data with the rest of the scientific community, even though, for now, they are obtaining a 'free ride' by not paying for their space on the station.

Dr Clark thinks that part of this tension 'comes down to the thinking that "my stuff is better."' Everybody knows there is going to be a limited amount of space on the station, so if a commercial venture can get into space faster, or easier, the scientists say, 'my stuff's not going to get on because of those people, who don't have to go through peer review, don't have to publish their data – and they get on board.'

Dr Uri, Dr Clark, and the management people at NASA generally, do not consider that private interests competing with the scientific community is the major object of concern in the commercial programme. At least for a while, the problem is going to be coaxing companies to take the risk of going into space. Dr Uri expects there to be constraints during the assembly sequence, because 'we're building a new vehicle. I don't think we know everything there is to know about how to do that. There is going to be a lot of uncertainty in having regular access to space. And these commercial guys, that's what they really want. They do not want to spend two or three years planning for a project before it even launches. To them, that costs money. They'd like to come in as late as possible to the [payload] integration flow, and get launched.'

As a matter of fact, Dr Uri stated that 'right now may not be the ideal time for all commercial customers, because not all the processes are necessarily lined up to respond quickly to bring a commercial experiment on board at fairly late notice. I envision later on, when we're pretty much through the assembly, and we know how

Figure 7.16. Many of the life sciences experiments will require hands-on operation by the crew on the space station. They will require the kind of care that Dr Haig Keshishian is giving these fruit fly larvae in a petri dish, and that were part of an experiment to examine the effects of microgravity on the development of neural connections between motor neurons and their targets in muscle fibres. The experiment flew on the STS-93 mission, which also deployed the Chandra X-ray Observatory to Earth orbit.

the processes have to work, and we've streamlined those and become better at it and more efficient, it will be a better opportunity. There will also be more time for the crew to operate those kind of things.'

He wanted to make clear, however, that he did not mean that commercial payloads should be excluded in the early phases of the station. 'Some are willing to take that chance', although it is a risk to rely on any particular flight. 'We've gone through several changes in the assembly sequence in the last couple of years,' he said, 'and that is likely to continue until we have more pieces of the station on orbit. All of that carries some risk. You could plan to fly a payload next June, and then all of a sudden you realise it's not next June, it's next December. And that costs money. I think some of the commercial folks are a bit leery about signing up until those things settle out.'

Dr Clark adds that the commercial sector's activity 'is really an applied science, more than it is a fundamental science. Yes, they will learn some fundamental scientific principles, but that's not really the idea of *most* of what is the commercial sector's task. They're really interested, for example, in protein crystal growth, of getting a better crystal faster, or growing fewer crystals in order to get a better one. The reason why they haven't paid cash money to do it is because we at NASA have not given enough access to space to show them that there is a bottom line, that it will save them that much money. It's hard to overcome all of the different obstacles in order to get to space.'

Responding to the criticism that NASA is providing free rides for the private sector, Dr Clark replied: 'We're never going to commercialise any of the space station if we don't give the store away a little bit to get people interested, because people don't understand the possible uses.' Never one to mince words, she continued: 'The commercial sector is not patient. Scientists will wait for a really long time, but most of the scientists who are complaining that they don't get to fly, don't get paid by NASA to come to work every day. They get their grant money, even if they don't fly. You get grant money for a long time. Nobody in the commercial sector gets paid to do that.'

While there are people coming forward from 'industries' such as entertainment, who 'think there may be a way to make some really big dollars, a lot of the science-based interests say it costs so much to get there,' she said. 'It's not free, and we don't really give it away, contrary to what the scientists are happy to say. It's not necessarily worth it to industry, so far. Hopefully, access to space in the form of the station will help solve that problem.'

One problem alluded to by Dr Clark is that of defining what NASA means by 'commercialisation'. Unlike the science experiments, the commercial sector will pay money to NASA, and could potentially fly anything in space, even if it is frivolous. 'I will let you know what I'm putting there is safe, and will not cause a problem to the station. Other than that, it's pretty much none of your business what I'm putting in there', can be one attitude of the commercial participants. There is a real issue, then, she said, that this experiment is 'taking up space from some real experiment'. Perhaps the space agency does not want to take this approach. NASA will have to define what it means by 'commercialisation.'

It is possible that conflicts may develop between contending national space agencies as to the very definition of 'commercial'. At the meeting of the Ministerial Council of ESA, in Brussels in 1999, the decision was taken to investigate the possible commercialisation of the European part of the ISS. One possibility under consideration is the choosing of a Commercial Agent, responsible for marketing and sales of ISS commercial utilisation for the space agency. The 1 October 1999 issue of the *LowG Journal*, published in Germany, reports that in addition to the traditional sectors in industry that could become interested in using the space environment for the development of new products, other 'businesses' could also become engaged. In this, they include promotion, public relations, advertising and marketing.

For forty years, NASA has refused to allow any of its facilities (or even photographs) to be used for advertising purposes. The space agency, and most members of Congress and citizens of the country, consider that NASA's space infrastructure is a national investment, dedicated to the furtherance of science, technology, applications, and exploration for 'the benefit of all mankind'. However, the Russians, especially when their financial situation became tenuous, have had cosmonauts 'advertise' various products from Mir. There will undoubtedly have to be some international 'rules' established as to whether, or to what extent, the ISS will be available for paid entertainment, advertisements, and other non-traditional 'business' uses.

'At some point', Dr Clark counsels, 'we're going to just have to learn to share the station, and potentially share information with each other, while we figure out how to maintain the proprietary nature for the commercial sector.'

7.8 FILLING THE SCIENCE GAP

From the moment that President Ronald Reagan uttered the words in his State of the Union address in January 1984, instructing NASA to build a space station within a decade, there has been a continual tug-of-war with the established institutions representing the scientific community, to garner their support and engage their participation in the project.

Similar to the opposition to the Apollo lunar landing programme of President John F. Kennedy, rather than seeing this new 'ocean of space' as an opportunity and challenge to conduct scientific research in a wholly new environment, many preferred to complain that the funding for this endeavour would only drain funds from their own treasured research projects. While it has been proven time and time again that when funding is cut for one area of science and technology it is also most likely to be cut in others, this fight continues to the present day.

One of the loudest voices against the International Space Station is that of Dr Robert Park, the director of the Washington office of the American Physical Society, who faxes a one-page weekly newsletter to the press. Although there is no indication that he has ever been involved in space research, or even bothered to find out very much about it, he pretends to speak for the scientific community in print, and at every opportunity. For example: 'Most scientists regard microgravity as one of the least important variables you can have', he told *Popular Science* in May 1998.

Dr Kathy Clark has a few choice words for Dr Park and the other self-appointed emissaries of the scientific community. She observes that 'NASA is in a very funny situation, because they are so well known, and they are beloved, and the only government agency that is, but it also makes them a target for all kinds of problems.' Recently, she reported, 'science writer Timothy Ferris wrote an article about how nothing has ever come from the space programme except Tang and jogging bras. He said that there is unanimous consensus among scientists that there is no scientific use for the space station... It's a weird thing that a person like Timothy Ferris, who is a writer and not a scientist, can stand up and make this statement... Then there's Robert Park,' she continued. 'I got interviewed by CNN, and I was talking to the reporter, and he said something about a scientist saying that there wasn't going to be any science. I said, 'That would be Robert Park.' And he said, 'Yes.' I said, 'Can't you broaden your horizons a little bit. Find another scientist. Robert is clearly making a living from saying this. He's never been to space, he doesn't know anything about it.''

The complaints have become louder as the cost of designing, building, testing and deploying the space station has increased over time. But advancements, as any

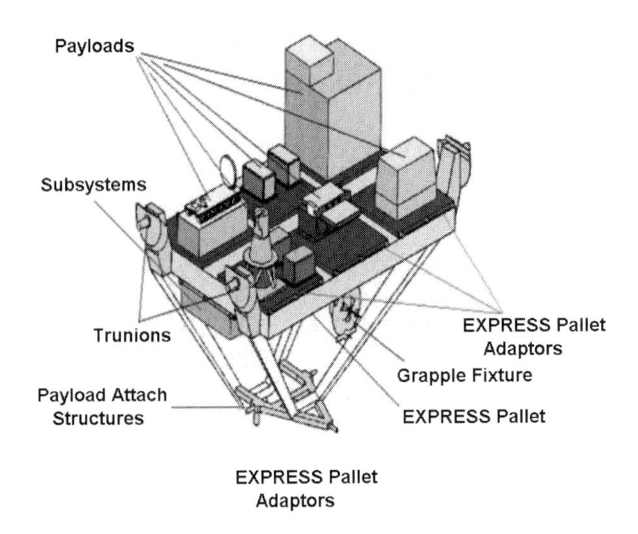

Figure 7.17. Brazil is the first developing nation to participate in the International Space Station. One of its contributions will be an EXPRESS (Expedite the Processing of Experiments to Space Station) rack, which will be placed outside the station on the truss structure in the Japanese Experiment Module Exposed Facility. The crew will use robotic systems for installation and removal of the attached payloads, without the necessity of space-walks outside the station.

scientist should know, cannot always be precisely budgeted and timelined. While it has taken longer, and cost more, to build the ISS than originally planned, this is partly because it is subject not only to delays and budget caps and cuts from the US, but also from 15 other partners. The financial problems in Russia that have led to delays are well known. The Canadian government had to renegotiate its agreements with NASA in 1994, when politicians there tried to cancel Canada's participation in the ISS and its contribution of the robotic Mobile Servicing Center, due to funding problems.

More recently, the space agency of Brazil asked for NASA's help in meeting its ISS commitments due to reduced funding. Following the signing of a bilateral agreement with the US in 1997, Brazil committed to provide NASA with a Technology Experiment Facility and EXPRESS pallet, to accommodate experiments external to the station. They are also to supply a Window Observational Research Facility, to be used as a mount with data and power connections, for optical experiments in Earth observation. Now, NASA is trying to plan for alternatives if Brazil cannot meet its obligations.

The delays in deploying the segments of the ISS have stretched out the schedule for starting many of the experiments that have been planned. NASA has instead been forced to redistribute funding from the future science programmes to the ISS engineering and fabrication activities. In 1997, NASA presented to the National Research Council in Washington (input is important for involving the scientific

Figure 7.18. In return for its contribution of hardware to the ISS, Brazil will have an astronaut on the station. Marco C. Pontes, a Major in the Brazilian Air Force, reported for training at the Johnson Space Center in August 1998. He will receive technical assignments within the JSC Astronaut Office, and then be assigned to a space station flight.

community in planning science on the ISS) a revised science research programme plan that 'borrowed' monies slated for science facilities in order to build the basic station hardware. This was necessary because the Clinton Administration and the US Congress had foolishly agreed to cap the annual funding for the ISS at $2.1 billion, due to pressures on the federal budget, not taking into account the difficulties the project might encounter along the way.

In a letter to NASA Administrator Goldin dated 8 July 1997, officials from the National Research Council stated that one of the 'challenges currently facing the space station programme [is] that imposed by its flat development funding profile'. The Space Studies Board of the NRC states that they 'understand that circumstances beyond NASA's control' have created many of the difficulties in the present situation. The letter summarizes the two to three-year delays expected in orbiting many of the research facilities for the ISS, and remarks that while space on Shuttle flights for small experiments will be available, this does not obviate the need for the more capable ISS facilities. The board members express concern that this delay could damage the momentum in the scientific community to participate in the ISS, and recommends that 'there may be opportunitites in the 1999–2000 time-frame to dedicate some Shuttle missions to life sciences and microgravity research.'

I asked Dr Uri how NASA was planning to accommodate the need for continuity in the space science research. He explained that 'if you go back maybe two or three years, we hadn't planned to do any significant research for a while on the ISS. We have the utilisation flights, and the first one is UF-1, which is currently scheduled for November of 2000 [as of August 1999]. We really hadn't planned to do anything before that first mission, and even that was to be fairly modest. But we realised that if we have some free time or other resources, we should try to get some research done earlier in the assembly phase.' As a result, 'there's been this backwards creep, closer and closer to the start of the assembly, to get some research done. It's not always an easy road, because there are a lot of constraints.' Because the science will have to take a back seat to the assembly tasks, there had not been any plan for science early in the assembly phase, 'so I think we're getting better at it than we originally thought we were.'

'As far as the hiatus,' Dr Uri said, 'that's going to be one of those facts of life, due to budgetary constraints, given that the budgets we get from Congress can't support flying additional Spacelab flights on the Space Shuttle and building the ISS at the same time. Something had to give, and if we want to build the ISS, we've got to get around to doing that.' NASA could plan to fly more science missions on one of the Space Shuttle orbiters, he reported. '*Columbia* is the one vehicle that does not go to the station because it's heavier, so there are opportunities every once in a while, perhaps, to fly one of those gap-filler science missions on *Columbia*. You've got to have the funding. [But] the House Appropriations Committee wanted to slash NASA's Fiscal Year 2000 budget by $1.4 billion', which was later reversed by the Senate.

Dr Clark concurs with the approach NASA is taking. The agency is trying to 'start some research on the station before it is done,' she stated. 'If we were going to just build the station first and then start research, we'd have built the station in a different way. Because we were worried about a hiatus in the science, we're actually constructing the station in such a way that research can begin almost immediately after the US laboratory goes up. It won't be full-fledged, and it won't be all the different disciplines, but you'll be able to do some things.' One of these was to move up the ability to deploy the small EXPRESS racks to hold experiments.

Dr Clark explained that one of the problems now is that the laboratory, Destiny, is behind schedule, because it is no longer just a research facility, 'but also the control of the station, with all of the gyroscope control and avionics in that lab'. It has been a complicated job to configure all of the hardware and software, she explained. 'We weren't going to do that to begin with, and we've ended up gaining a lot, but it's been a very difficult process. We keep finding problems, and it sounds like a terrible thing that we're delaying launches, but imagine trying to fix these problems on orbit!'

This is not the first time there will be a gap in space science research, Dr Uri recalled. 'If you think back, in 1974 when Skylab ended, and 1983 when we flew our first Spacelab mission [in the Space Shuttle], we weren't doing a whole heck of a lot of research in space, either. Not in human spaceflight. From 1983 to the last Spacelab which flew last year, that was a great period of time when we got a lot of science done. Now, we're taking another dip because the station got delayed. ISS was supposed to be up and running by the time we were finishing up with the Neurolab mission last year, so there would not have been such a gap. Mir came along, and that gave us a whole different flavour of what kind of science you can do on long-duration missions. Then our commitment to Mir ended, and ISS kept slipping out, so that gap kind of formed itself. We've tried to minimise its impact by doing

Figure 7.19. In order to carry out scientific research on the International Space Station before fully-fledged standard payload racks are available, NASA is planning to deploy EXPRESS (Expedite the Processing of Experiments to Space Station) racks in the Destiny laboratory. The racks have already been flight-tested in the Space Shuttle. In July 1997, Michael Gernhardt observed experiments in the Astro/Plant Generic Bioprocessing Apparatus, set in the EXPRESS rack on the Space Shuttle *Columbia*.

research as early as possible on the ISS. Hopefully, five years from now we'll look back and say 'we got through that drought, and we survived.'

In attempting to help close the gap between the end of the dedicated Space Shuttle science missions and research on the ISS, NASA has solicited experiments to be designed for early use on the station even before the Destiny laboratory is deployed. At the end of 1998, the space station office at NASA called for experiments that could operate before there is a full-time expedition crew on the ISS, and which can operate unattended for about two months. Two of the most likely candidates will be the protein crystal growth experiments (discussed in Chapter 5) – the Enhanced Dewar and the Diffusion-Controlled Apparatus for Microgravity (DCAM), both of which operated successfully on Mir. These experiments would be conducted in an EXPRESS rack designed to handle experiments with minimal complexity, and could even be placed inside the Unity module or Zarya, both of which are already in orbit, or on the Russian Service Module, Zvezda.

Even though the delays in deploying research facilities has afforded the scientific community more time to develop and choose experiments for the ISS, *Science* magazine's David Malakoff complains that 'only' 100 US experiments have been approved for flight. But this is largely due to the fact that the machinery is still being developed to have the proposals from each nation be subjected to international peer review, which will then produce the array of flight experiments. Malakoff himself admits that the limited number of experiments approved so far is *not* a function of disinterest. Japanese scientists, he reports, submitted more than 750 proposals in two rounds of requests, with about 50 still under consideration for launch. In 1997, European researchers submitted nearly 100 proposals for experiments to be externally attached to the station, and the first international call for life sciences experiments 'attracted more than 500 proposals from the United States, Canada, Japan, and Europe'. Scientists already engaged in space science research, or those who are at least open to considering the advantages, are excited about the opportunities of the ISS.

7.9 FOR ALL MANKIND

There are 16 nations that are partners in the ISS: Belgium, Brazil, Canada, Denmark, France, Germany, Italy, Japan, The Netherlands, Norway, Russia, Spain, Sweden, Switzerland, the United Kingdom, and the United States. There are more than 100,000 individuals involved in the ISS at space agencies around the world, and under hundreds of contract and subcontractor companies. There is no question that this is already a global programme.

On 20 July 1969, when American Apollo astronauts landed on the Moon, they left a plaque there that stated that they had come in peace 'for all mankind'. The International Space Station, for the first time, provides 'all mankind' with the opportunity to actually participate in this great adventure. Brazil is the first newly-industrialising nation to join this manned space project. As the space station reaches full operation, more developing nations should consider how to become part of the space age.

Scientists from all nations should submit proposals to the international partners, to involve their countries in research aboard the station. One of the greatest contributions space exploration makes to 'all mankind' is the inspiration it instills in young people around the globe. The more countries that are involved, the broader will be the impact on young scientists.

The ISS will be an evolving structure. Even when its reaches 'assembly complete' in 2004, its modular design will enable new capabilities to be added. The ISS could become a check-out garage for vehicles heading out far from Earth. It could be a testbed for advanced propulsion and other technologies, that cannot be adequately tested on Earth. The station could be the site for *in situ* tests of life support and other systems that will be required to live on the Moon and Mars.

Hardware that has not yet even been conceived might be developed by other nations over the first decade of the century. Any such contribution to the space station would be 'in kind', with no money changing hands. Brazil has been given the opportunity to fly an astronaut on the station, in exchange for the payloads it has committed to deliver. Other nations could take advantage of that opening to send one of their citizens into space.

Figure 7.20. The International Space Station is the most complex engineering task that has ever been undertaken, not only because of the technology and science, but because of the involvement of fifteen nations. On 29 January 1998, senior government officials from all of the participating countries signed agreements in Washington DC to establish the framework of the design, development, operation and utilisation of the space station. Some of these officials then accompanied NASA Administrator Dan Goldin (front, sixth from the left) on a tour of the Kennedy Space Center. The appropriately named Unity node is in the background.

What is holding many nations back from participating in this current phase of human space exploration, and is slowing the pace of what even the international partners can accomplish on the station, is their being unable to allocate the funds. This is not a problem of absolute resources, but of priority. There are surely the resources in Russia to meet all of its commitments to the ISS. There is a vast scientific and technical capability in the former Soviet republics that is no longer utilised. The design bureaux and aerospace production factories that created a decade of 'firsts' when the Space Age began, and then a succession of space stations, still stand. But Russia today is in the throes of an economic collapse, brought on by its adherence to the policies of international financial institutions, which have contracted the real economy, thrown thousands of scientists and engineers out of work, and instead deployed monetary resources in financial speculation and crime.

American economist Lyndon LaRouche Jr has stated in absolute terms that the space programme does not 'cost' anything. The investments made in visionary projects and aerospace technologies create the wealth in the economy that the society can then deploy to meet its needs on Earth, and lay the foundation for growth for future generations. No one today asks how much money it cost to build the railroads, or provide electricity to towns and factories. Those investments, made possible by government policies to promote their growth, laid the basis for our economic strength over the last two centuries.

In monetary terms, one can calculate that if scientific developments on the space station were to lead to more effective treatments that substantially reduce the cost of treating diabetes, or preventing the suffering from widespread infectious diseases such as influenza, the space station will have paid for itself, in dollars, many times over.

In 1957, at the dawn of the Space Age, Krafft Ehricke, a German–American space pioneer and the quintessential philosopher of the Space Age, developed the concepts he believed should guide space exploration. In his 'Anthropology of Astronautics', he proposed three laws for this new science of 'operating in space and travelling to other worlds'. The first law is that 'nobody and nothing under the natural laws of this Universe impose any limitations on man except man himself.' Second, 'not only the Earth, but the entire Solar System, and as much of the Universe as he can reach under the laws of nature, are man's rightful field of activity'. The third law states: that 'by expanding through the Universe, man fulfils his destiny as an element of life, endowed with the power of reason and the wisdom of the moral law within himself.' Dr Ehricke would later describe these laws as constituting an 'extraterrestrial imperative' for mankind.

In 1974 Dr Ehricke presented an invited lecture at the 25th Congress of the International Astronautical Federation in The Netherlands, entitled, 'Space Stations – Tools of New Growth in an Open World'. 'Space stations', he said, 'like all technical means, are tools that receive meaning only through the goals and objectives to which they are put. I believe our space objectives should, above all, contribute to the preservation and growth of modern, enlightened civilisation.'

Dr Ehricke's second point was that 'human development – the central purpose of civilisation – cannot be pursued without growth. If we look at our world today as having reached the limits to growth we should have every cause to despair'. He often stressed in his writing that 'fortunately, there are no recognisable limits to growth. The course of evolution over billions of years, overcoming each and every limit to growth at that time, attests to this.' In our time, the exploration and colonisation of space is the gateway to the new 'open world'.

Does that mean that, automatically, mankind will chose the right path at this juncture? Dr Arthur Kantrowitz, a physicist who has been involved in many innovate lines of research throughout his 50-year career in science, and who presented a concept for a four-man space station with Krafft Ehricke in 1958, recounted in *21st Century Science & Technology* in 1990, how in the fifteenth century, the Ming Dynasty in China had led a flowering of science and technology, culminating in fleets of 1,500-ton ships. In 1420, these 'treasure ships' made their way down the east coast of Africa, preceding by 50 years the Portuguese voyages down the west coast of Africa.

China made its last great sea voyage in 1431–1433. In 1436, when a new emperor came to the throne, he issued an edict that forbade the building of ships for overseas voyages. There were reports that the fleets were burned. This policy led to a decline

Figure 7.21. The International Space Station is just the beginning. It will provide the basis for travels to distant worlds which have been the dream of mankind ever since he looked up at the sky and gazed at the Moon, the planets and the stars. (Concept by Krafft Ehricke; illustration by Christopher Sloan.)

in many branches of science and technology, a dampening of the spirit of the Chinese people, and an isolation which eventually allowed the invasion of the nation four centuries later during the Opium Wars. Apparently, in the view of 'right-thinking bureaucrats', the Grand Fleet of Treasure Ships consumed funds which could have been better spent on domestic problems, such as water conservation and farming.

Dr Kantrowitz retold this story of the Ming Dynasty ten years ago, due to his concern that this same retreat from science and technology was occurring in the United States, as well as in other nations of the world. We have made a transition, he said, away from 'an adventurous technology policy in favour of a proof-of-'safety' philosophy of life. This was done in spite of a general awareness', he says, 'that the adventurous policy had been brilliantly successful in the past in improving our health and safety, and in spite of awareness that *a priori* proof of safety is never possible.'

In 1957 Dr Ehricke could already see that 'the concept of space travel carries with it enormous impact, because it challenges man on practically all fronts of his physical and spiritual existence. The idea of travelling to other celestial bodies reflects to the highest degree the independence and agility of the human mind. It lends ultimate dignity to man's technical and scientific endeavours. Above all,' he wrote, 'it touches on the philosophy of his very existence. As a result, the concept of space travel disregards national borders, refuses to recognise differences of historical or ethnological origin, and penetrates the fibre of one sociological or political creed as fast as that of the next.' Space travel, he believed, is a universal activity that redefines man's relationship to the rest of the Universe.

Considering the hardships and courage it has taken for man to explore other frontiers such as the North Pole and the Antarctic, there is little doubt that the International Space Station programme will be a very high-risk venture, on a path to the even greater risks and rewards of living on the Moon and travelling to Mars.

Throughout history, visionaries and explorers have been willing to take up the challenge. It is only the faint-hearted who stay at home.

Appendix 1

American long-duration missions on Mir

Mission	Astronaut	Delivery vehicle	Launch date	Return vehicle	Landing date	Days on Mir
NASA 1	Norman Thagard	Soyuz 70	1995 Mar 14	STS-71	1995 Jul 7	111
NASA 2	Shannon Lucid	STS-76	1996 Mar 22	STS-79	1996 Sep 26	184
NASA 3	John Blaha	STS-79	1996 Sep 16	STS-81	1997 Jan 22	123
NASA 4	Jerry Linenger	STS-81	1997 Jan 12	STS-84	1997 May 24	127
NASA 5	Michael Foale	STS-84	1997 May 15	STS-86	1997 Oct 7	138
NASA 6	David Wolf	STS-86	1997 Sep 26	STS-89	1998 Feb 1	124
NASA 7	Andrew Thomas	STS-89	1998 Jan 23	STS-91	1998 Jun 12	13

Appendix 2

Psychological support of American astronauts on Mir

Astronaut	Family/Friend		Conference with guests	News and entertainment programmes
	TV	Telephone		
Norman Thagard	3	3	3	5
Shannon Lucid	8	17	13	8
John Blaha	5	10	7	7
Jerry Linenger	3	14	5	6
Michael Foale	2	14	12	28
David Wolf	6	14	9	19
Andrew Thomas	5	9	13	7

Appendix 3

Scientific investigations on Mir

Discipline	NASA long-duration increment						
	1	2	3	4	5	6	7
Advanced Technology	–	1	2	–	–	1	3
Earth Sciences	–	2	2	2	3	3	3
Fundamental Biology	1	3	2	4	5	1	–
Human Life Sciences	26	11	12	8	6	5	6
ISS Risk Mitigation	–	5	7	8	7	6	2
Microgravity	1	12	10	11	9	9	8
Space Sciences	–	2	2	2	–	–	–
Total	28	36	37	35	30	25	22

Bibliography

Ames Research Center Press Release. 'US, Russia Plan Collaborative Study of Plant Growth in Space', 13 May 1997.

Ames Research Center Press Release. 'Scientists Announce Germination of Seeds Grown in Space', 1 August 1997.

Associated Press. 'A Veteran Is Longing to Return to Space', 21 September 1999.

Augsburger, Judy. Russians Reluctant to Bid Mir Adieu, *MSNBC*, 26 August 1999.

Baevsky, R., Nikulina, G. and Funtova, I. 'Development of Medical Forecasting Methods on International Space Station: Research of Cardiovascular Regulation during Long Space Flights.' Presented at the Congress of the International Astronautical Federation, Amsterdam. October 1999.

Becker, Jeanne. Testimony. House Committee on Science, Space, and Technology, Subcommittee on Space, 9 March 1993.

Becker, Jeanne. Testimony. House Committee on Science, Space, and Technology, Subcommittee on Space, 22 June 1993.

Becker, Jeanne. Women's Health Issues and Space-Based Medical Technologies, *Earth Space Review*, **3**, no.2, 1994.

Belew, Leland, (*ed.*). *Skylab: Our First Space Station*. NASA S-400, 1977.

Belew, Leland F. and Stuhlinger, Ernst. *Skylab: A Guidebook*. NASA E-107, 1973.

Binckmann, E. and Brillouet, C.A. 'Biology in Space: ESA's Biolan and Modular Cultivation System on the ISS.' Presented at the Congress of the International Astronautical Federation, Amsterdam, October 1999.

Bingham, G.E. *et al.* 'Microgravity Effects on Water Supply and Substrate Properties in Porous Matrix Root Support Systems.' Presented at the Congress of the International Astronautical Federation, Turin, Italy, October 1997.

Bingham, G.E. Interview, 5 August 1999.

Bioreactor: Ground-Based Research Results. http://science.msfc.nasa.gov/newhome/br/ground.htm, (Undated).

Brown, David. NASA Ready to Attempt Human Cell Growth in Space. *Aviation Week & Space Technology*. **137**, no.1, 6 July 1992.

Brown, Irene. 'Seven Americans Called Mir Home'. space.com, 27 August 1999.

Campbell, William F. *et al.* 'Lessons Learned from Greenhouse-Integrated Wheat

Experiments in Svet on Mir.' Presented at the Third Phase I Research Program Results Symposium, Huntsville, Alabama, November 1998.
Carter, Daniel C. Interview, 12 August 1999.
Carter, Daniel C. *et al.* Diffusion-Controlled Crystallization Apparatus for Microgravity (DCAM): Flight and Ground-Based Applications. *Journal of Crystal Growth*, **196**, 1999, 602–609.
Carter, Daniel C. *et al.* PCAM: A Multi-User Facility-Based Protein Crystallization Apparatus for Microgravity. *Journal of Crystal Growth*, **196**, 1999, 610–622.
Carter, Daniel C. *et al.* Lower Dimer Impurity Incorporation may result in Higher Perfection of HEWL Crystals Grown in Microgravity: A Case Study. *Journal of Crystal Growth*. **196**, 1999, 623–637.
Cauquil, D. *et al.* 'Cardiolab: A French–German Contribution to the European Physiological Modules Facility in Columbus.' Presented at the Congress of the International Astronautical Federation, Amsterdam, October 1999.
Churchill, Suzanne. *Fundamentals of Space Life Sciences*. Krieger Publishing Company, Malabar, Florida 1997.
Clark, Kathy. Interview, 26 August 1999.
Clark, Phillip. *The Soviet Manned Space Program*. Orion Books, New York, 1988.
Compton, W.D. and Benson, C.D. *Living and Working in Space: A History of Skylab*. NASA SP-4208. 1983.
Connors, M., Harrison, A. and Akins, F. *Living Aloft: Human Requirements for Extended Spaceflight*. NASA SP-483. 1985.
Cooper, Henry S.F., Jr. *A House in Space*. Holt, Rinehart and Winston, New York, 1976.
Covault, Craig. Lucid Describes 'Fantastic' Voyage. *Aviation Week*, **145**, no.14, 30 September 1996.
Cowling, Keith. Science x Hype = Bad Policy + No Benefit. *NASA Watch*, 20 May 1999.
David, Leonard. International Space Station. *Aerospace America*, **37**, no.7, July 1999.
Day, John, and McPherson, Alexander. Macromolecular Crystal Growth Experiments on International Microgravity Laboratory-1. *Protein Science*, **1**, 1992, 1254–1268.
DeLucas, Lawrence J., *et al.* Protein Crystal Growth in Microgravity. *Science*, **246**, 3 November 1989.
DeLucas, Lawrence J. Testimony. House Committee on Science, Space and Technology, 22 June 1993.
DeLucas, Lawrence J., and McDonald, William. 180-Day Operation Accomplishments Report, X-Ray Detector Test (XDT) Experiment. 10 March 1999. Unpublished.
DeLucas, Lawrence J. *et al.* 'The International Space Station X-Ray Crystallography Facility.' Presentation at the 48th Annual Denver X-Ray Conference, 2–6 August 1999.
DiChristina, Mariette. Weird Science: Research on the Space Station is Really Out There. *Popular Science*, May 1998.

Duray, P., Hatfill, S. and Pellis, N. Tissue Culture in Microgravity. *Science & Medicine*, **4**, no.3. May/June 1997.

Ehricke, Krafft. 'Space Stations – Tools of New Growth in an Open World.' Presented at the Congress of the International Astronautical Federation, Amsterdam, October 1974.

Engle, Eloise, and Lott, Arnold. *Man in Flight*. Leeward Publications Inc, Annapolis, Maryland, 1979.

Fisher, David. Shannon Lucid's 188 Days in Space: From STS-76 to STS-79. *Quest*, **5**, no.3, 1996.

Freed, L., Langer, R., Martin, I., Pellis, N. and Vunjak-Novakovic, G. Tissue Engineering of Cartilage in Space. *Proceedings of the National Academy of Sciences*. **94**, December 1997, 13,885–13,890.

Freeman, Marsha. The United States Must Save Soviet Science. *Executive Intelligence Review*. **19**, no.18, 1 May 1992.

Freeman, Marsha. *How We Got to the Moon: The Story of the German Space Pioneers*. 21st Century Science Associates, Washington DC, 1993.

Freeman, Marsha. Krafft Ehricke's Extraterrestrial Imperative. *21st Century Science & Technology*, **7**, no.4, Winter 1994.

Freeman, Marsha. U.S. and Russia Link Manned Space Programs. *Executive Intelligence Review*, **22**, no.31, 4 August 1995.

Freeman, Marsha. Shuttle–Mir Program is Under Threat from Congress. *Executive Intelligence Review*, **24**, no.40, 3 October 1997.

Freeman, Marsha. The Mir Space Station: A Technological Achievement. *Executive Intelligence Review*, **24**, no.41, 10 October 1997.

Freeman, Marsha. Space Research on the Medical Frontier: Interview with Dr David Wolf. *Executive Intelligence Review*, **25**, no.11, 13 March 1998.

Freeman, Marsha. The Mir Space Station: Man's Courage to Explore. *21st Century Science & Technology*, **12**, no.1, Spring 1999.

Freiberg, P. Psychology Keeps Astronauts Well Grounded: Structures that Psychologists Put in Place are Critical to Astronauts' Success. *Journal of the Society for Human Performance in Extreme Environments*, **4**, no.1, April 1999.

Gargir, G., Guell, A. and Zappoli, B. 'The French Mir Missions and Preparations for Use of the International Space Station.' Presented at the Congress of the International Astronautical Federation, Amsterdam. October 1999.

Gazenko, O.G. Psychological Aspects of Operating Large Cooperative Space Ventures Involving Multi-National Crews. *Orbital International Laboratory*, American Astronautical Society, **33**, Science and Technology. 1974.

Goodwin, T., Prewitt, T., Spaulding, G. and Becker, J. Three-Dimensional Culture of a Mixed Mullerian Tumor of the Ovary: Expression of *In Vivo* Characteristics. *In Vitro Cell Developmental Biology*, **33**, May 1997, 366–374.

Goodwin, Thomas. Interviews, 29 July and 23 August 1999.

Grigoriev, A., Kozlovskaya, I. and Kozerenko, O. 'Multicultural Joint System of Medical Support for ISS: What Can Be Learned From Russian Long Term Space Flights.' Presented at the Congress of the International Astronautical Federation, Amsterdam, October 1999.

Hammond, Timothy *et al.* Gene Expression in Space. *Nature Medicine*, **5**, no.4, April 1999.

Hammond, Timothy. Interviews, 13 and 23 August 1999.

Hammond Timothy *et al.* 'Genetic Analysis of Mechanisms of Tissue Engineering.' Submitted for publication.

Harland, David M. *The Mir Space Station: A Precursor to Space Colonization.* Wiley–Praxis, Chichester, England, 1997.

Ivanova, Tania *et al.* 'The First "Space" Vegetables Have Been Grown in the "SVET" Greenhouse Using Controlled Environmental Conditions.' Presented at the Congress of the International Astronautical Federation, Montreal, Canada, October 1991.

Ivanova, Tania. 'SVET' Biotechnological System, Controlling the Environmental Conditions for Growing Higher Plants in Weightlessness.' Presented at the Congress of the International Astronautical Federation, Washington DC, August 1992.

Ivanova, Tania, Kostov, P. and Sapunova, S. 'Renewing of the 'Greenhouse SVET' Biotechnological Experiments Onboard the Mir Space Station.' Presented at the Congress of the International Astronautical Federation, Oslo, Norway, October 1995.

Ivanova, Tania *et al.* 'Earth Activities Referring to Improvement of Soil Moistening Control of Space Experiments in SVET Space Greenhouse.' Presented at the Congress of the International Astronautical Federation, Turin, Italy, October 1997.

Ivanova, Tania. Private communication, 8 September 1999.

Jessup, J. Milburn. Testimony. House Committee on Science, Space, and Technology, Subcommittee on Space, 22 June 1993.

Jessup, J. Milburn. Interview, 24 September 1997.

Johnson, Nicholas. *Soviet Space Programs: 1980–1985.* Science and Technology Series, Vol.66, American Astronautical Society, 1987.

Johnston, Richard. Skylab Medical Program Overview. *Proceedings of the Skylab Life Sciences Symposium, August 27–29, 1974.* NASA TM X-58154.

Johnston, Richard, and Dietlein, Lawrence. *Biomedical Results from Skylab.* NASA SP-377, 1977.

Kanas, N. Psychiatric Issues Affecting Long-Duration Space Missions. *Aviation, Space, and Environmental Medicine*, **69**, no.12, December 1998.

Kanas, N., Weiss, D. and Marmar, C. Crewmember Interactions During a Mir Space Station Simulation. *Aviation, Space, and Environmental Medicine*, **67**, no.10, October 1996.

Kanas, N. *et al.* 'Social and Cultural Issues During Shuttle/Mir Space Missions.' Presented at the Congress of the International Astronautical Federation, Amsterdam, October 1999.

Kantrowitz, Arthur. The U.S. Space Program and the Ming Navy. *21st Century Science & Technology*, **3**, no.2, Spring 1990.

Kelly, A. and Kanas, N. Crewmember Communication in Space: A Survey of Astronauts and Cosmonauts. *Aviation, Space, and Environmental Medicine*, **63**, no.8, August 1992.

Kelly, A. and Kanas, N. Communication between Space Crews and Ground Personnel: A Survey of Astronauts and Cosmonauts. *Aviation, Space, and Environmental Medicine*, **64**, no.9, September 1993.

Kelly, A. and Kanas, N. Leisure Time Activities in Space: A Survey of Astronauts and Cosmonauts. *Acta Astronautica*, **32**, no.6. 1994.

Key, Charles. *The Story of 20th Century Exploration*. Alfred A. Knopf Inc, New York, 1937.

Kidger, Neville. NASA Science Comes to Mir. *Spaceflight*, **38**, September 1996.

Koszelak, Stanley *et al.* Protein and Virus Crystal Growth on International Microgravity Laboratory-2. *Biophysical Journal*, **69**, July 1995, 13–19.

Koszelak, Stanley, Leja, Cathy, and McPherson, Alexander. Crystallization of Biological Macromolecules from Flash Frozen Samples on the Russian Space Station Mir. *Biotechnology and Bioengineering*, **52**, 1996, 449–459.

Koszelak, Stanley. Interview, 3 August 1999.

Kozerenko, P. *et al.* Some Problems of Group Interaction in Prolonged Space Flights. *Journal of the Society for Human Performance in Extreme Environments*, **4**, no.1, April 1999.

Kozlovskaya, I.B. 'Evaluation of Efficacy of Physical Exercises as a Countermeasure Means in Long Term Space Flights. 'Presented at the Congress of the International Astronautical Federation, Amsterdam, October 1999.

Kozlovskaya, I.B. *et al.* 'Countermeasures Study on Board the International Space Station Russian Service Module.' Presented at the Congress of the International Astronautical Federation, Amsterdam, October 1999.

Kroes, Roger. Interview, 9 August 1999.

Langer, Robert, and Vacanti, Joseph. Tissue Engineering: The Challenges Ahead. *Scientific American*, **280**, no.4, April 1999.

Lebedev, Oleg. 'Selection of Crewmembers for a Unique 240-Day Experiment on Flight Simulation Onboard ISS to begin in September.' RIA–Novosti, 22 July 1998.

Lebedev, Valentin. *Diary of a Cosmonaut: 211 Days in Space*. Phytoresource Research Inc, Texas, 1988.

Lipman, Jonathan. 'Kathy Clark: A Kid in Her Stellar Sandbox.' space.com, 25 October 1999.

Littke, Walter, and John, Christina. Protein Single Crystal Growth Under Microgravity. *Science*, **225**, 13 July 1984.

Long, Marianna *et al.* Protein Crystal Growth in Microgravity, Review of Large-Scale Temperature Induction Method: Bovine Insulin, Human Insulin, and Human Alpha Interferon. *Journal of Crystal Growth*, **168**, 1996, 233–243.

Longhurst, Frank. 'Status and Outlook of European Participation to the ISS.' Presented at the Congress of the International Astronautical Federation, Amsterdam, October 1999.

LowG Journal. ISS – Research Centre and Marketplace. 3 October 1999.

Lucid, Shannon. Six Months on Mir. *Scientific American*, **278**, no.15, May 1998.

Lucid, Shannon. Interview, 12 August 1999.

Malakoff, David. $100 Billion Orbiting Lab Takes Shape. What Will It Do? *Science*, **284**, 14 May 1999.

Manzey, D. and Lorenz, B. Human Performance During Spaceflight. *Journal of the Society for Human in Extreme Environments*, **4**, no.1, April 1999.

McDonald, William. Interview, 9 August 1999.

McIntire, Larry. Testimony. House Committee on Science, Space, and Technology, Subcommittee on Space, 22 June 1993.

McPherson, Alexander. Macromolecular Crystals. *Scientific American*, **260**, no.3, March 1989.

McPherson, Alexander. Virus and Protein Crystal Growth on Earth and in Microgravity. *Journal of Physics D: Applied Physics*, **26**, 1993, B104–B112.

McPherson, Alexander. Recent Advances in the Microgravity Crystallization of Biological Macromolecules. *Trends in Biotechnology*, **15**, no.6, June 1997.

McPherson, Alexander. Interview, 17 August 1999.

McPherson, Alexander *et al.* The Effects of Microgravity on Protein Crystallization: Evidence for Concentration Gradients Around Growing Crystals. *Journal of Crystal Growth*, **196**, 1999, 572–586.

McPherson, Alexander *et al.* 'ISS: A Science Classroom for America.' 1999, unpublished.

McPherson, Alexander *et al.* An Observable Protein Crystal Growth Apparatus for Studying the Effects of Microgravity on Protein Crystallization. *Proceedings of the Space Technology & Applications International Forum–2000*. American Institute of Physics, New York. In press.

Meyer, Patrick, and Bray, Becky. Interview with Shannon Lucid. *Liftoff*, Marshall Space Flight Center, March 1998.

Microgravity News. Welcome Home: The First Microgravity Experiment Returns from Mir. Spring 1996.

Microgravity News. Winter 1996.

Microgravity News. ISS: Continuing the Journey. Spring 1998.

Microgravity News. Engineering the Crystal. Spring 1998.

Microgravity News. Culturing a Future. Autumn 1998.

Microgravity News. Conference on ISS Utilization. Spring 1999.

Microgravity News. Lisa Freed and Gordana Vunjak-Novakovic: Profile. Autumn 1998.

Microgravity News. Looking to the Crystal for Answers. Summer 1999.

Mooney, David, and Mikos, Antonios. Growing New Organs. *Scientific American*, **280**, no.4, April 1999.

Moore, Karen, Long, Marianna, and DeLucas, Lawrence. Protein Crystal Growth in Microgravity: Status and Commercial Implications. *Proceedings of the Space Technology and Applications International Forum–1999*. American Institute of Physics, New York, 1999.

Moore, Karen. Interview, 25 August 1999.

Morphew, M. Ephimia, *et al.* Voyage of Discovery: American Astronauts Aboard Russia's Mir Space Station. *Journal of the Society for Human Performance in Extreme Environments*, **2**, no.1, June 1997.

Morphew, M. Ephimia. Personal communication, 14 September 1999.

Musgrave, Mary. Private communication, 2 September 1998.

NASA. 'NASA Joins the Fight Against Diabetes.' 9 June 1997.

NASA. On Target for a Cure. *NASA Science News*, 11 July 1997.

NASA Press Release. 'Space Station Technology Will Bring Expert Medical Care to Remote areas on Earth.' 12 September 1997.

NASA. 'The Phase 1 Program: The United States Prepares for the International Space Station.' 1998.

NASA. 'Space Grown Insulin Crystals Provide New Data on Diabetes.' 3 June 1998.

NASA. 'Space Station Will Put Experimenters 'On The Rack'.' 13 July 1998.

NASA. Weightless Research Has Heavy Implications. *NASA Science News*, 22 July 1998.

NASA. Using Space Incubator to Understand Breast Cancer. *Space Science News*, 1 October 1998.

NASA. Press release, 98–211, 23 October 1998.

NASA. Breathing Easier, Living Longer are Goals of Shuttle Experiments. *Space Science News*. 3 November 1998.

NASA. 'The International Space Station: Improving Life on Earth and in Space – The NASA Research Plan.' 1998.

NASA. 'The Phase 1 Program: The United States Prepares for the International Space Station.' 1998.

NASA. Laboratory Under Construction. *Space Science News*, 10 December 1998.

NASA. NASA Selects New Biotechnology Experiments for Development. *Space Science News*, 8 March 1999.

NASA. 'International Space Station Fact Book.' July 1999.

NASA. Room to Let; Easy Access; All Utilities. *Space Science News*, 14 September 1999.

NASA. Materials Science 101. *Space Science News*, 15 September 1999.

NASA. Scientists Grow Heart Tissue in Bioreactor. *Space Science News*, 5 October 1999.

NASA/NIH Center for Three-Dimensional Tissue Culture. Annual Report, February 1997.

NASA/RSA. Science & Technology Advisory Council Research; Final Research Reports. September 1997.

National Research Council. Letter to NASA Administrator Daniel Goldin, 8 July 1997.

Natsuisaka, M. and Yoda, S. 'NASDA Activitites on Microgravity Sciences for JEM Utilization.' Presented at the Congress of the International Astronautical Federation, Melbourne, October 1998.

Naumann, Robert. Interview, 5 August 1999.

Nechaev, A., Myasnikov, V. Stepanova, S. and Kozerenko, O. 'Psychophysiological State and Work–Rest Schedule Tensity as Factors Affecting Crew Members' Professional Reliability.' Presented at the Congress of the International Astronautical Federation, Amsterdam, October 1999.

Ng, Joseph *et al.* Comparative Analysis of Thaumatin Crystals Grown on Earth and in Microgravity. *Acta Crystallographica*, 1997, 724–733.

Nicogossian, Arnauld, and Parker, James. *Space Physiology and Medicine*. NASA SP-447, 1982.

Nicogossian, Arnauld (*ed.*). *Space Physiology and Medicine*, second edition. Lea & Febiger, Philadelphia, 1989.

Nicogossian, A., Pool, S., Wear, M. and Hamm, P. 'Updates in Long-Term Follow-up of Astronaut Health.' Presented at the Congress of the International Astronautical Federation, Amsterdam, October 1999.

Oberth, Hermann. *Die Rakete zu den Planetenraumen* [*Rockets Into Planetary Space*]. R. Oldenbourgh, Munich, 1923.

Pellis, Neal. Testimony. House Committee on Science, Space, and Technology, Subcommittee on Space, 22 June 1993.

Pellis, Neal. Testimony, House Committee on Science, Subcommittee on Space and Aeronautics, 10 April 1997.

Pellis, Neal. 'Neal Pellis discusses Thomas' Research Program.' NASA, Mir 25 Report, 13 March 1998.

Pellis, Neal. Interviews, 25 July 1997 and 9 August 1999.

Penley, N., Uri, J., Sivils, T. and Bartoe, J. 'Integration of Multiple Research Disciplines on the International Space Station.' Presented at the Congress of the International Astronautical Federation, Amsterdam, October 1999.

Phase 1. *The Phase 1 Program: The United States Prepares for the International Space Station*. NASA Monograph, 1999.

Pitts, John, A. *The Human Factor: Biomedicine in the Manned Space Program to 1980*. NASA SP-4213, 1985.

Rayl, A.J.S. Microgravity and Gene Expression: Early Results Point to Relationship. *The Scientist*, **13**, no.18, 13 September 1999.

Robey, Judith. 'NASA's International Microgravity Strategic Planning for the Space Station Era.' Presented at the Congress of the International Astronautical Federation, Amsterdam, October 1999.

Salmon, Andy. Science Onboard the Mir Space Station: 1984–1994. *Journal of the British Interplanetary Society*, **50**, 1997, 283–295.

Saradzhyan, Simon. Russian-Developed Lenses May Alter Cosmonaut's Mood. *Space News*, 19 July 1999.

Sarver, George, Hanel, Robert, and Givens, John. NASA's Gravitational Biology Research Laboratory for the International Space Station. *Proceedings of the Space Technology & Applications International Forum–1999*. American Institute of Physics, New York, 1999.

Sawyer, Kathy. Growing Spare Parts For People in Space? *The Washington Post*, 18 November 1991.

Scharp, David. Testimony. House Committee on Science, Space, and Technology, Subcommittee on Space, 22 June 1993.

Schneider, William, and Disher, John. Skylab Contributions to the Future. In *Space Stations Present and Future*, L.G. Napolitano (*ed.*), Pergamon Press, New York, 1976.

Silber, Kenneth. 'Mir Scientists Win Aerospace Prize.' space.com, 8 October 1999.

Snyder, Robert. Interview, 16 August 1999.

STS-70 Status Report no.11. NASA, 18 July 1995.

Stuhlinger, Ernst. Materials Processing Under Zero Gravity: An Assessment After

Skylab. In *Proceedings, Third Space Processing Symposium, Skylab Results, April 30–May 1, 1974*, Vol 1. Marshall Space Flight Center.

Stuster, Jack. *Bold Endeavors: Lessons from Polar and Space Exploration*. Naval Institute Press, Annapolis, Maryland, 1996.

Thomas, Andy. Interview, 10 December 1998.

Uri, John. Interview, 24 August 1999.

Unsworth, Brian, and Lelkes, Peter. Growing Tissues in Microgravity. *Nature Medicine*, **4**, no.8, August 1998.

Utkin, V. *et al*. 'Main Summary of Efforts on Structuring the Program on Research and Experiments of Start-Up Period on the ISS Russian Segment.' Presented at the Congress of the International Astronautical Federation, Amsterdam, October 1999.

Vernikos, Joan. 'Life Sciences Research on Space Station.' Presented at the Congress of the International Astronautical Federation, Amsterdam, October 1999.

Williams, R., Nicogossian, A. and Shepanek, M. 'NASA and Multilateral Medical Operations: Space Medicine in the ISS Era.' Presented at the Congress of the International Astronautical Federation, Amsterdam, October 1999.

Williams, David, and Flynn, Christopher. 'Cross-Cultural Considerations for Long-Duration Space Flight.' Presented at the Congress of the International Astronautical Federation, Amsterdam, October 1999.

Witt, A.F. *et al*. Steady State Growth and Segregation Under Zero G: InSb.' In *Proceedings, Third Space Processing Symposium, Skylab Results, April 30–May 1, 1974*, Vol 1. Marshall Space Flight Center.

Index